Signature 1994

Production Design for Television

Production Design for Television

Terry Byrne

Focal Press
Boston   London

Focal Press is an imprint of Butterworth–Heinemann.

Copyright © 1993 by Butterworth–Heinemann, a division of Reed Publishing (USA) Inc. All rights reserved.

No part of this publication may be reproduced, stored in a retrieval system, or transmitted in any form or by any means, electronic, mechanical, photocopying, recording, or otherwise, without the prior written permission of the publisher.

♾ Recognizing the importance of preserving what has been written, it is the policy of Butterworth–Heinemann to have books it publishes printed on acid-free paper, and we exert our best efforts to that end.

Library of Congress Cataloging in Publication Data
Byrne, Terry
    Production design for television. / Terry Byrne.
      p. cm.
    Includes index.
    ISBN 0-240-80120-2
    1. Television—Art direction.   I. Title
    PN1992.8.A76B97   1993
    791.45'023—dc20

British Library Cataloguing-in-Publication Data
A catalogue record for this book is available from the British Library.

Butterworth–Heinemann
80 Montvale Avenue
Stoneham, MA 02180

10 9 8 7 6 5 4 3 2 1

Printed in the United States of America

Contents

Preface ix
Acknowledgments xi

1 Introduction 1
What's a Production Designer? 1
The Designer's Roles 1

2 Origins 3
The Need to Be Entertained 3
Formats 3
Talent 6
Formula 6
Stage to Film 6
Film to Television 8
Beginning with Theatre 11
Architectural Influences 13
Traditional Art Influences 14
Training 15
New Directions 15

3 Roles and Responsibilities 17
Staff 17
Production Design 18
Costume Design 20
Lighting Design 22

4 Objectives 24
The Bottom Line 24
The Commercial Role of Design 24
Design and Advertising 25
Clarity in Objectives 27

5 Composition   31
Controlling the Focus   31
Talking Heads   32
Graphics   33
Multiple-Head Shots   34

6 Focus   37
Directing Focus   37
Single Camera or Multiple Cameras   41

7 Dynamics   44
The Use of Motion   44
Designing Backgrounds: Putting Motion in Context   44
Moving Cameras   45
Dynamic Texture   46

8 Color   48
The History of Color   48
Designing in Color   49
Color, Lighting, and Camera Technology   51
The Psychology of Color   52
Paints and Pigments   54

9 Mechanics   56
The Film Model   56
The Videotape Recording System   56
Advantages of Electronic Imaging   58
Special Effects   59
The Impact of Computer Technology   59
Digital Advances   61
What All This Means to the Designers   61
Assembling the Product   61

10 Interpretation   64
Researching the Interpretation   64
Developing the Design Solution   65
Lateral Thinking   67
Inherent Interpretation   68
Tools at Hand   68
Getting It Right   69

Ongoing Input 70
Some Good Examples 70

11 Distortion and Fantasy 73
Degrees of Interpretation 73
The Mechanics of Fantasy 74
The Machinery 77
Animation 81

12 Script and Storyboard 85
Building on the Verbal Foundation 85
Analysis 89
Storyboards 91

13 Rendering and Developing the Idea 97
Rendering as a Process 97
Preliminary Rendering Techniques 97
Evolution of the Idea 99
Model-Making 100
Rendering on Location 101
The Final Rendering 103

14 Plots and Plans 106
Documentation Drawings 106
Floor Plans and Site Plans 106
Elevations and Details 108
Plots 110
Drafting Technique 115
Media 115
Language 118
Style 123

15 Sets and Props 125
Working Conditions 125
Building 126
Flats 127
Platforms 133
Stairs 135
Special Construction 137
Furniture and Props 139

Finishing   140
3-D Treatments   140
Painted Treatments   141
Set Dressing   144

16 Makeup and Costumes   147
Roles and Responsibilities   147
Costume Organization   148
Makeup Organization   150
Summary   152

17 Lighting   154
Roles and Responsibilities   154
Rudimentary Lighting Principles   156
Key and Fill   157
Backlight   158
Dressing Light   159
Lighting Mechanics   161
Lighting Aesthetics   166
Summary   169

18 Graphics   171
Parameters   171
Personnel   172
Artwork   172
Aesthetics   174

19 Where Do We Go from Here?   176
Interactivity   176
Outside the Mainstream: Nonbroadcast Television   178
Further Technical Advances   179
Where the Designer Fits In   179

Glossary   181
Recommended Reading   187
Index   191

Preface

Production design for the theatre has long been taught in colleges, art schools, and theatre-training programs, and many fine books about stage design and lighting have been written over the years.

Although employment prospects in legitimate theatre have steadily declined, those courses continue to be taught and new texts continue to be published. I felt it was important to try to address the subject of production design from a perspective more consistent with my own experience and, I believe, more consistent with real-world practice. That experience has led me into the corporate theatre, facility consulting, and television industries, and has taught me that the skills I learned in the study of theatre design could easily be adapted to other purposes and would serve me well in any number of related applications. Those discoveries are at the core of my approach in this book.

I have also been fortunate enough to have worked in both the United States and Europe (an experience reflecting another way in which the world is changing) and thus have tried to make this book adaptable to both markets. Our colleagues are increasingly crossing borders in pursuit of new and challenging design work, and the result is a much greater degree of stimulation and cross-pollination—the better for us all. I would encourage anyone who has not worked abroad to pursue any opportunities to do so.

This text assumes a basic background in art and some exposure—even if only very elemental—to theatre, film, or television practice (The subject is a broad one, and I do intend that the serious student should acquire *all* of the books listed in the Recommended Readings, and more.) It would be difficult indeed to master production design without a solid comprehension of the process of script interpretation, directing, studio organization, construction practice, and the fundamentals of design and draftsmanship. Anyone seriously determined to pursue a career in design for television should understand this and set out to learn the fundamentals. Hopefully, college curricula will soon begin to address this subject and will make it possible to study television production design in an integrated and sequentially constructed program.

Acknowledgments

This book is the product of the experience, wisdom, and encouragement of many people. I was fortunate to have had some excellent teachers and colleagues along the way, and my deepest thanks go to all of them—especially to Oren Parker and Pete Ralph for being there and for being generously supportive at pivotal junctures.

A number of people have made generous contributions of time, knowledge, and materials without which this book would be far leaner. My many friends and associates at Radio Telefis Eireann in Dublin deserve special mention—especially Alan Farquharson, whose work is represented in several chapters. Ian Rawnsley, Graham Shepherd, and others at the BBC in London have also been exceptionally helpful and generous. Frank Schneider at WNBC in New York deserves special mention for his patience, generosity, and expertise, as do Adrianne Kerchner at KYW-TV in Philadelphia, Joe Wood, John Gresch, and John Wells in Los Angeles, and Cletus and Barbara Anderson at Carnegie-Mellon University.

More material support was provided by Ron Weiss at Korey, Kay & Partners in New York; Benita Silverberg at BBC/Lionheart Television; Steve O'Driscoll and Cory Resh at Modern Video Productions in Philadelphia; Stephen Chamberlain at Arriflex Corporation; and Kelly Murphy at Quantel Corporation.

Introduction 1

What's a Production Designer?

Some years ago, I was making small talk at a party and answered the inevitable question about my line of work by explaining that I was a production designer. Faced with the usual perplexed frown, I went on to explain, "a television set designer." The very polite lady with whom I was talking then asked, "Do you mean the boxes or the bits that go inside?"

Although there are many of us around, production designers seem to be a well-camouflaged breed; many didn't study to become what we are, and most got into the business by accident. Frank Schneider, who was one of the originals and may be the last staff designer at the New York studios of the National Broadcasting Company (NBC), studied business in college and sort of fell into a job with the network—painting scenery. Shortly after joining the network, he admitted he was hooked, and he went back to school to study theatre and, finally to become a production designer.

I went to college and then to graduate school to become a theatre set designer (on Broadway, of course). After several years of pounding on that door and numerous way-off-Broadway jobs, I also fell into television. Apart from the heady, dizzying feeling induced by a regular income, I can honestly say that I've never—before or since—had so much fun and gotten paid for it.

The production designer is in an enviable position. She is a creator—one who has a lot of input at the early, formative stages of program development—and also a manager, trusted with a budget and a great many support services. The impact of the production designer is widespread, and the opportunities to interact with all the different types of personalities indigenous to a production is exhilarating. (Sometimes it's also exasperating, but the business is like that.) I highly recommend such a career to those with the will to pursue it to fruition and something valuable to contribute.

The Designer's Roles

If a designer were simply another variation of artist, the job would be a simple one. As much appeal as art has, it is somewhat self-involved; the painting or sculpture is the whole thing, and the artist's job is largely finished when that piece is done. Designers, on the other hand, are collaborators in a larger process. Whether graphic design, interior design, or production design, it is a process that depends for its success on the efforts and craft skills of many others to reach its own fruition. The sketches and drawings, no matter how well executed, are merely the first part of a longer sequence of events and a preliminary expression of a much more complex and dynamic idea. The thrill is in seeing one's work on screen; seeing how it interacts with the performers or even becomes a performer in its own right. No painting in the world can excite me as much as the sight of a set of my own creation seen under studio lights for the first time, with cameras moving around and through it, and performers making it come alive.

To accomplish this, we must be artists, architects, costume experts, lighting experts, model-makers, and dabblers in a hundred other crafts and trades as dictated by the specifics of a given project. It doesn't sound boring, does it? The responsibilities can be enormous. Many is the night I've wakened at some ungodly hour in a cold sweat imagining all the things that might go wrong with an ambitious load-in or a complicated shoot day and had to get by on three or four hours of sleep, having spent the night making checklists and checking them again and again until I dozed off.

The designer is often blamed for others' mistakes and is frequently called on to fix the unfixable or to do some other impossible task. In general, this means that your skills at improvisation; clear analytical thought; tact; and, finally, bluntness will be called into play at one time or another.

Nonetheless, may I reiterate that seeing one's masterpiece realized under lights and on-screen and hearing a producer's accolade (such as, "It looks nice, really.") is actually *worth the effort*.

Origins 2

The Need to Be Entertained

It is often mistakenly assumed that the histories of the various performing arts have somehow evolved in isolation from each other. I recall a short piece presented on television, encapsulating the history of the Mack Sennett comedies in which the speaker, a man well versed in film history, listed the items of visual slapstick "invented" by Sennett. No credit was given to nor mention made of the actors of the Commedia Dell'Arte, who performed the same visual gags hundreds of years before the Keystone Kops. In the same way, analysts of television are inclined to think that history began in the late 1940s, ignoring the immense debt owed to vaudeville and early film in the shaping of form and establishment of a vocabulary and a framework within which the dialogue between television and its audience could take place.

As long as we have been social animals (probably as long as we have been identifiably human), some of us have felt a need to perform and others of us have felt inclined to be entertained. This is elemental; the need for such social interaction, the need to enact our fantasies and realize our imaginings has driven us to construct increasingly elaborate frameworks within which to entertain each other Only the limitations of the technology current at a given time have restricted the elaborateness of these frameworks.

Formats

The Greek theatres, by and large, were carved from stone, although many scholars believe that these stone structures were complemented with wooden structures approximating stage scenery and a fairly elaborate mechanical construction that allowed for the *deus ex machina* and other special effects not a great deal different in mechanism or effect from the flying harness still being used for Peter Pan. The old tricks are often still the best.

Let's take a historical leap into the Renaissance, when the advent of opera and the popularity among royalty of masques and command performances brought about a need for grand scenic effects. Remember that the stage established a context in which the audience could suspend its disbelief and vicariously experience the events happening to the fictional characters of the play within their fictional world. The borrowing of sailors' techniques for rolling and unrolling huge expanses of canvas on which entire scenes of buildings, landscapes, and seascapes were painted permitted

Figure 2.1 A Renaissance stage. After Inigo Jones

changes of location to be effected within minutes. At that time, the theatre moved indoors and was lit with candles (and, later, gas or oil lamps), removing the context of performance still further from reality. The burden of controlling focus (in this usage, the focus of the audience's attention) began to move away from the actor and onto the scenery and lighting, a trend that would accelerate into the twentieth century.

Such scenery and lighting, being necessarily absent from the outdoor drama of the Greeks and Medieval Europe, did not enter into the equation in any important way until the Renaissance. Prior to that time, broad gesture, costumes, masks, and acoustic amplification of the actors' voices were the only means of emphasis and support available. For comic effect, caricature masks and phalluses were used and for dramatic effect more important characters were given elevator shoes and shoulder pads. In the comedy of the Commedia Dell'Arte, the characters were categorized by type, each wearing a standardized costume that identified him to the audience and established audience's expectations of him. All of these devices established the context within which the action of the performance worked and they defined a set of conventions (tacitly agreed-upon terms of communication between performers and audience) that permitted the audience to accept the action onstage as fantasy and yet feel involved in that action enough to enjoy the humor and feel sadness at the tragedy.

At that time, performing outdoors on a hillside or in a village square, the burden of commanding the audience's attention was entirely borne by the cast. Technology has eased that burden extensively. Since the age of candlelight, with the coming of electric lighting and the adoption in the theatre of advanced optics, we have acquired highly directional and focusable lighting equipment, which enables the lighting designer to throw a pinpoint beam of light onto the stage from as far away as 200 feet. It is thus no longer necessary (although still effective in some comedy) to equip actors with exaggerated body parts to hold the audience's attention. Having acquired absolute control over the illumination of the performing area, we can control focus simply by lighting only that which we want to be seen. A musical comedy with a stage full of dancers will be lit in broad strokes; banks of spotlights wash the stage at a very high level of luminance and demand the attention of the spectators simply by virtue of volume and intensity. At the other extreme, an intense emotional scene with only two actors onstage can be isolated within a tight pool of light and the audience made to forget momentarily the rest of the stage and the 2000 other people in the room with them.

The next step in this progression was the invention of the cinema. While the advance of stage-lighting technology enabled theatre practitioners to control focus to a degree inconceivable to their counterparts of Shakespeare's time, it is still true that, for most of the audience, the emotional two-handed scene in the spotlight is over 50 feet away and, therefore, relatively small and hard to see. We can control an audience's attention by restricting the light, but we cannot enlarge that which is distant or very small. No matter how sophisticated theatre technology becomes, the actors must still rely on relatively broad gestures to convey often subtle emotions. Film is different; the camera can go as close as the eye can go, or closer. Not only can it control our focus to an infinitely greater degree than is possible onstage, but it can also magnify subtle details which cannot possibly be perceived over distance by the unaided eye.

In terms of scenery, too, the advent of film made huge advances possible. A stage play happens, of necessity, in real time: While it is possible to perform a play with flashbacks and to jump from one time frame to another, wholesale changes of setting and abrupt juxtapositions of visuals are at best time-consuming, noisy, and cumbersome. At worst, they are distracting and ineffectual. The fact that film is not a real-time medium—that the performance is removed in time from the production by weeks, months, or years—buys the film maker some options which are not available to the stage director. Film can fade from one image to another, it can superimpose images on one another, and it can cut abruptly from bright daylight to pitch-black night. The camera can move around its subject as a theatre audience cannot; it can create images of worlds that don't exist, animate inanimate objects, and make dreams come to life before our eyes. While the magic of theatre is the tension of real-time performance, the presence in the same room of the characters of the play, the magic of film is its ability to do what we, living in the real world in real time, cannot. Theatre makes things larger than life in order to draw our attention to them; film takes us places we can't normally go in real circumstances in order to achieve the same result. Remember that the goals of both are essentially the same: Both media do their work by telling stories. They capture our attention and help us, through our imagination, to experience another person's vision vicariously and thus to broaden our own vision. The more effectively they do this, the more skillfully their practitioners use the tools of each medium, the better they work.

Talent

The skills of stage actors are, obviously, different from those of film actors. The need to memorize two hours' worth of lines, rather than ten minutes', and the need to make gestures and voices larger than real in order to carry over distance are two important differences. For the director and the designer, understanding compositional tools and stage dynamics helps to control emphasis and make the event onstage, though a magnification from life, remain consistent within that context and so take on a reality of its own. Movement of actors and setting elements, dynamics of lighting changes, and traffic-control decisions, for example, have to be carefully made in order to build properly—to support the goals of the piece rather than detract from them.

Formula

The design of the stage set must use certain accepted formulas of composition (I firmly believe there are no rules other than cause and effect) so as to avoid overwhelming the action, distracting from the action, or proving inadequate to the visual or practical necessities. This involves detailed and protracted consideration of sight lines from various points in the auditorium to the principal points of action onstage in order that the composition of the stage space works equally well for the majority of the audience. This is the single greatest obstacle to the stage designer—small budgets can be worked with, artistic differences can be resolved with patience, but bad sight lines can only be resolved by remodeling or demolishing the theatre. There are few, if any, Broadway theatres in which the audience at the extreme sides of the hall cannot see behind the set to the stagehands standing in the wings. The filmmaker who doesn't want something on the set to be seen merely changes the angle of the shot.

Stage to Film

The lessons of composition, visual control, and rhythm learned from the theatre are often useful to the directors and designers of films. Not every shot can be a close-up; for the simple reason of variety (among others) it is necessary to change types of shots, and the composition of the elements within each shot must be considered and planned in the same way that the design of a stage set and the movement of the actors within it is planned for the theatre. In fact, due to the enlargement of everything the camera sees, and thus the exaggeration of film, greater emphasis is placed on the details of composition, because the significance of each detail is also exaggerated.

The style of acting for the camera, then, tends to reinforce the small things. The tiny movement of facial muscles, or subtle changes in vocal tenor, for example, are made bigger than life and given greater significance. Because of the piecework nature of the film assembly process, the need to memorize long passages of dialogue is removed. In fact, the importance of the word is often reduced. Because the audience in the theatre cannot see small changes of facial expression, many subtleties must be conveyed vocally; whereas a small movement can convey volumes on camera.

For the designer, the differences between theatre and film are somewhat more complex. If some members of the theatre audience are seated very near the set, the level of detail in terms of stage properties and general finish of the set pieces must be quite fine. Nonetheless, the overall scale of the set and the arrangement of its elements, must have the desired effect on those audience members who are seated at the back of the theatre. A crucial part of this is the creation of an environment for the audience as well as for the characters of the play; it is vital to assist the audience in feeling itself to be in the same environment with the performers in order to encourage that essential suspension of disbelief. In the 1960s and 70s, environmental settings, such as those done by Jerry Rojo, tried to abolish any distance in the minds of the audience between themselves and the performance. They went to the extent of building a somewhat fanciful, constructivist slave ship, for example, out of scaffolding, and making the audience sit in parts of the hold with the actors playing slaves. Eugene and Franne Lee (since occupied extensively with design for television) did a set for *Candide* on Broadway, which placed the action all over the theatre and put the orchestra under the balcony, all in an attempt to break the conceptual barrier between performers and audience represented by that invisible fourth wall of the proscenium opening.

What has this to do with film? Remember that film and theatre are competing for the same audiences: those who go out to an auditorium (in its broadest sense, this can apply to a theatre or a cinema) to participate in a group social experience involving the enactment of a story in which they will be vicarious participants. (The other big competitor for the public's entertainment time is television, but it doesn't compete on the same terms because it is not really a social phenomenon in this sense.) Film has the advantage: A theatre set can be mammoth and elaborate and a very intricate machine, but it must repeat its mechanical feats every night and twice on Sundays. A film set need only work once. Thus, the theatre is competing in terms of creating elaborate and fanciful environments for an audience with an increasingly voracious appetite for special effects.

Since *Star Wars* and thanks largely to people like George Lucas, an industry devoted entirely to special effects has sprung up in Los Angeles and London. In addition, the use of elaborate traveling matte compositions in features is increasing. The film industry is thus more inclined to produce films with elaborate effects, able to do them well and relatively inexpensively, and able to deliver the final product to its audience for a fraction of the cost of a production of similar impact in the live theatre. For these reasons, there has been an increase in science-fiction film production.

Film to Television

As mentioned previously, film and theatre have historically been separate from the television market. Although early in its history the film industry felt television to be a threat to its security (Warner Brothers would not, for many years, permit a TV to be seen in one of its film sets in the hope that what is not seen will cease to exist), it soon became apparent that television provided a very lucrative aftermarket for feature films and could increase, rather than diminish, the profits of the film production houses. Thus, the film companies got into the business of producing product for television in the late 1950s, and the lines of demarcation between film and television became indistinct.

From the outset, television drew much of its talent and its ideas about presentation from the theatre rather than from film. It is important to remember that television was regarded as a live medium: The videotape recorder became viable well after the advent of broadcast TV, so most early programming went out live. In addition, early television was black and white and early cameras were of low resolution and high contrast, producing a flat and very unsubtle monochrome image. These factors, along with the small screen, virtually insured that TV was no competition for the feature film.

Television did assume the roles of several film products besides features, however. Television news replaced the Saturday newsreel and its Saturday adventure serials usurped the serial Westerns shown at the movie

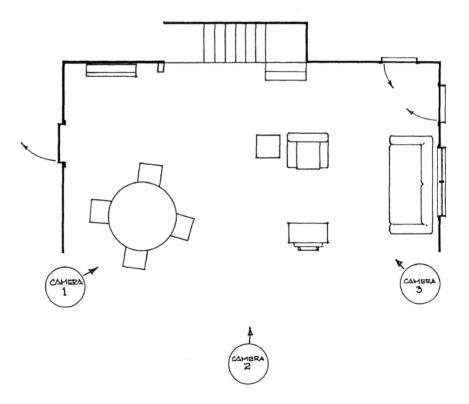

Figure 2.2 A sitcom box set

houses in the 1930s and 1940s. However, because the film production companies quickly assumed the role of producer of these TV serials, they suffered little financial loss, if any.

Because TV was a live medium, many of the performers came from theatre and radio. In general, the variety and comedy artists came from vaudeville or burlesque and the news people came from radio. Milton Berle, Lucille Ball, and Red Skelton came from music halls and nightclubs, and did little to change their acts for television. Because editing scenes together was not an option in the early years, TV shows had to be shot with multiple cameras, and usually had live audiences to supply reaction sound effects. The arrangement that worked most efficiently was one in which the actors played against a long, shallow set, open on one side for the three or four cameras and the audience (if there was one) behind the camera positions. In effect, television simply lifted the old proscenium arrangement from the theatre, complete with the missing fourth wall and dropped it into a television studio, replacing the audience (wholly or in part) with cameras. In this setup—which still predominates in the situation comedy—the shots tend to be flat, since all the cameras are lined up on one side of the set, and the performers (of necessity) all tend to line up "downstage center" facing the same direction, just as they did in the theatre of the late nineteenth century.

Audiences have always been willing to meet their entertainers halfway in suspending their disbelief and allowing their imaginations to fill in the blank spots, but never more so than when the end result is a laugh. This hasn't changed; the scenery in most sitcoms is rather badly painted—the level of detail anything but realistic—and the furnishings and props only those that are actually used in the action of a typical episode. The same is also true of soap operas, but the audiences for these programs do actually get quite caught up in the stories and often refer to certain programs of this type as "realistic."

The soaps, or soap operas, are not incorrectly named. Their sponsors originally were soap companies, and some still are; Colgate-Palmolive continues to maintain its own production company for this purpose. Their plot lines are certainly no less preposterous than many an opera plot; they are rife with heroes and heroines, perplexing mazes of twisted subplots, and astounding improbabilities of hidden relationships and identities and even supernatural occurrences. These aspects of operatic and melodramatic theatre found a willing audience in television; it is inherently an escapist medium, figuring in most people's lives as a release from the tension of reality. In fact, opera was a predecessor in this, the music and the romantic plots combined with the elaborate spectacle to draw people away from the routine of their existences for awhile. Its successors—melodrama, operetta, and musical comedy—all owed their appeal to the same escapist need of their audiences, and, having established their places in society, paved the way for television to answer the same need in much the same way. From "The Honeymooners" through "Married With Children" the form (and formula) has remained almost perfectly intact, with broad characterizations, standup and slapstick gags, a low level of finish in terms of set treatment, and flat frontal lighting in a standard box set that could have been lifted from a stage play of a century ago. This is not to criticize, but merely to point out that the needs of the audience have not changed, and the means by which television meets those needs is not substantially different from the time of vaudeville or melodrama or opera.

There are many reasons why formulas used in the development of television programming do not change, but the primary reason is the driving force behind the industry: advertising sales. The bottom line in all commercial TV is advertising; a popular show gets viewers, which get ratings, which determine the cost of advertising slots, and thus directly regulate network profits. Thus it is that a given network, having come up with a formula for a particular sitcom that proves popular, is very reluctant to change it. Programming is regarded (and often referred to) as "product" rather than as a kind of artistic creation. In much the same way that Detroit develops a new model car, hypes it, and then proceeds to crank out as many as it thinks it can sell, Los Angeles develops TV programs and cranks them out as quickly and efficiently as possible. Unless the audience demands it, these creations tend to look very much like one another, varying in details of situation and being built around the particular idiosyncracies of different performers, but the fundamental aspects of structure and even the jokes are not very different from one another.

Some notable exceptions to this formula approach entered the market in the 1980s and 90s, and with some success. Series such as "Hill Street Blues" and "China Beach" have been built on a much more true-to-life style of characterization and more realistic production design and lighting borrowed from the high end of the film tradition rather than from vaudeville. These programs address a different audience need than the sitcoms, and are less escapist and less reliant on fantasy to attract audience.

Credit for the truce between the film and television industries is usually assigned to "Cheyenne," the first of the great television Westerns. Originally intended as a feature film, this was a project of Warner studios, which was sold to NBC in 1955 and began the stampede of production houses feeding the TV networks from New York to Hollywood. The network production facilities had been located in New York because the pool of talent (that is, stage and radio) on which the industry relied was already located there. After the great exodus to California happened, a number of changes took place. The range of performance styles enlarged to encompass dramatic series, especially drama in the film style rather than the theatrical, and there was a greater exchange of talent between the two media. It is no longer unusual to see performers freely jumping between television and film, and such actors as Michael J. Fox have begun their careers in television and easily moved from there to feature film stardom—a move once considered highly unlikely if not impossible.

Also important is the fact that most bigger budget programs originated in the U.S. television market are done on film rather than videotape. This is because of the substantial aftermarket sales overseas: Despite the fact that television has been with us for over 40 years, there is still no internationally standardized format. In the United States, we use a TV system called NTSC which was developed in the early 1950s, and is of somewhat poorer picture quality than the other predominant system, called PAL, which was developed in Europe. In order to distribute programs to other countries and get the best quality picture on the system that is standard there, it is easier to use film, a format that is standardized worldwide, and until the arrival of high-definition television (HDTV), was a higher-quality origination format than any videotape format.

The net result was that television adopted the film-production technique (which was already in place when all this began in 1955) for the making of its high-end drama specials and serials. In general, TV drama tends to use its cameras film-style rather than theatre-style as the sitcoms do. This means that, rather than using a proscenium-style setting and arranging the cameras along the "fourth wall," the entire scene is shot from one angle, then the camera is moved and the entire scene shot again from another angle, and so on until all of the camera angles required are complete. The raw footage is then edited, cut by cut, until the finished scene with the required changes of shot is ready. As you can imagine, this requires that the setting be re-dressed and re-lit for each take and that each performance be virtually identical to the others so that they will edit together neatly. The editing process is necessarily very time consuming and expensive. For those in the industry whose goal is a cheap and efficiently produced product, shooting film style is out of the question. However, there is an audience for such programs, and the growing market for U.S. programs abroad (together with an increasingly efficient worldwide distribution and marketing system for TV product) is making high-end production much more profitable. Of course, the TV Westerns that started it all, such as "Gunsmoke" and "Maverick," are still seen in reruns on both broadcast and cable channels, and the prospects are good that the same pattern will hold true for those series produced since. Certainly the continued success of "M*A*S*H" and "Hill Street Blues" in reruns would augur well.

Ongoing technical advances are also making high-end product easier and less expensive to produce. Film's greatest handicap is the delay time required between shooting and processing, which means that if a scene needs to be reshot the production crew may not know that until a day or so later. With videotape, of course, immediate playback is possible, so a scene can be reshot before anything on the set has been removed or changed in any way. The arrival of the Sony high-definition system, which shoots a video image in a filmlike format and with 1125 lines of resolution (as good as film) has provided another means of origination in a high-quality format and possesses all of the advantages of videotape.

Beginning with Theatre

During the 1960s and 70s, universities all over the United States had thriving theatre departments. Many students (myself included) were diligently engaged in studying acting, design, and technical theatre and fully expected to make careers in the business. Most were disappointed; they found the theatre industry to be crowded, rather poorly paid, and very competitive. Nonetheless, their training served them well; most found that the variety of skills they'd learned and the rough-and-ready adaptability of theatre practices made a shift to a parallel career path quite easy. Some who studied lighting design became architectural lighting designers, finding their theatrical flair and ability to use colored light very marketable in the architectural industry. Others found work in more closely related fields, such as the industrial show or the commercial exhibition industries. Many more found that there were more jobs available in the television industry, especially with the advent of cable and corporate television in the late 1970s.

We've looked at the theatrical predecessors to TV and film; it should be apparent by now that the skills acquired in the study and practice of theatre are readily translatable to television. Television borrowed many of its practices from the theatre (as well as from film), and many of its practitioners as well. One lighting designer studied theatre lighting in college and went to work for NBC in the early 1950s as a lighting director. In those days, the cameras were clumsy and not very sensitive and had only fixed-focus lenses with limited depth of field. The parameters of television lighting had been defined by laboratory technicians and, as such, dictated very high levels of intensity, sometimes 250 to 300 footcandles. Coming from a background that treated lighting as a creative medium rather than as a purely technical one, he experimented with levels—using his eye as his guide rather than relying completely on the light meter—and discovered that he could get much more realistic and three-dimensional pictures by varying levels and using less light than that for which the camera manufacturer's technical manual called. This is heresy to certain sorts of television engineers. This battle is ongoing, but the fact is that many lighting directors in the United States and elsewhere have been varying light levels down as low as 50 footcandles for certain effects and coming up with excellent results.

Those who learned their craft in the theatre typically began working at the poverty level, doing lighting or sets or costumes for small companies with microscopic budgets. They have, of necessity, learned to experiment with their equipment to find new uses for it, to adapt equipment from other industries to uses for theatre, and always to fix their minds upon a desired effect and to bend the tools and equipment at hand to that purpose. This willingness to flex, to experiment with equipment until the goal is achieved is the chief advantage theatre-trained people bring to television.

When I made the transition to TV design, I asked a friend who had done so some years before me for any tips he might care to pass along. (I had spent some time in the NBC studios at McDonald Avenue in Brooklyn and was awestruck by the amount of cables, the size of the cameras, and the control rooms full of colored lights and other technical paraphernalia.) His one piece of advice was, "Forget about sight lines." Sight lines, you'll recall, are the bane of a theatre designer's existence; he was saying, in essence, that the television designer has a great deal more control over visual images than the stage designer and faces fewer frustrations as a result.

Essentially, this is true. While it is necessary to adapt in many areas and learn some new technology in order to make that transition, the level of support for the designer and the control mechanisms available make for a somewhat more rewarding working environment. Cameras do not process colors in the same way that the human eye does, nor are they as sensitive; therefore, it is necessary to constantly refer back to a control monitor (having made certain that it was properly adjusted in the first place) to see whether the visual effect is coming across as it was supposed to. It is usually necessary to deal with design elements at a finer level of detail than in the theatre. The camera can see things closer in than can the theatre audience (especially since the 15-power zoom lens became standard) and thus shows flaws that would be overlooked onstage. Lighting becomes more difficult because there are more obstructions (cameras and audio equipment) in studio than onstage. These are far from insurmountable, and an individual with training and experience in theatre design can usually adapt to television practice without suffering any great amount of angst.

As is the case with many of the best actors, some designers also return to live performance frequently—not for the money, but to give themselves artistic variety and to sharpen their skills. Television design (especially at the network production level) involves a larger number of support personnel and is a complicated and often cumbersome machine. It is sometimes an exercise in retention of one's sanity to get back to the simpler process of stage design with its more fundamental challenges and more immediate rewards.

Architectural Influences

At the other end of the creative spectrum, but nonetheless closely allied to both theatre and television design, is the older art of the architect. Many of the great scenic artists and set designers were also architects. The famous Bibiena family, originators of classic Renaissance opera design, as well as the Englishman Inigo Jones and the American Norman Bel Geddes, were highly respected designers of stage settings and architects.

Again, the essential elements of the two forms are very closely allied: Much of stage design involves the representation onstage of buildings from real life (or those that could have existed in real life) and requires an understanding of the principles of composition, balance, and structure used by the architect in the design of a building. An essential component of any stage design curriculum is a course in the history of architectural style. The stage designer needs to be fluent in the language of the architect: the various types of columns and friezes used by the Greeks, the florid details of the rococo, down to the Russian or Norwegian variations of late Victorian eclectic architecture must all be familiar to the serious student of theatre set design.

More than simply memorizing the histories of different architectural styles, the designer must understand the social contexts within which they fit and from which they emerged. The people who lived in such spaces and their concepts of beauty must be understood, as should the function of those styles, so that the designer can adapt the style of a given time to make it fit the reality of the story. The designer must interpret in his or her own style (and a style supportive of the tenor of that specific production) while remaining true to the original style and inspiration of the period. Josef Svoboda, the great Czechoslovakian designer, was fond of assigning his students the task of designing one window for the set of a specific production of a specific play in such a way that anyone familiar with that play would know, when first seeing the window design, the production for which it was intended. Few of his students were fully successful at this.

The object of the exercise, though, was to stress the importance of the designer's immersion in the architecture of a given period, to the point of achieving an almost intuitive feel for it—at which point one can put away the reference books and just draw.

Anyone who has studied architecture comes to the task of set design with an advantage in instances where the understanding of architectural form is a critical part of that task. I remember a design professor accusing a graduate student (who had earned a bachelor's degree in architecture and was studying for his master's in stage design) of being constitutionally incapable of drawing a curved line (or anything other than a rectilinear composition). The professor was eventually proved wrong, but not without a good deal of hair pulling and general anguish from the student. The truth is that, although architectural forms often play a dominant role in set design, one must often have a command of other, less orderly sorts of composition.

The ability to sketch and render freehand is critical. If all human activity were contained indoors, the work of production design would be half as difficult; it isn't, however, and a command of the visual styles of nature is every bit as useful as that of man-made forms. At the low-budget end of television production, the scripts are contrived to keep the action indoors—but as the budgets rise to cover the additional costs of transporting a crew to locations outside the studio, the production designer has to become a landscape gardener, road builder, farmer, boatwright, or whatever may be required. Any of these may be approached in the same way that Svoboda approached the problem of the window—immersion in the art or craft that the story dictates until one is something of an expert. It is rarely sufficient just to hire an expert; a real boatwright, for example, may have a lifelong familiarity with his craft to draw upon, but he has no knowledge of the craft of the camera, and thus can't put his own handiwork into proper context for that camera. That is the job of the production designer, and the production designer must have a thorough knowledge and understanding of both in order to do the job well.

Traditional Art Influences

A number of working production designers have studied more traditional art subjects, such as painting, sculpture, or interior design. Fundamental drawing and rendering skills are essential, no matter what the formal training; it is no use having brilliant design concepts if you're not able to express them on paper to someone else. If a person is entering production design from one of the traditional art fields, the transition will require study of the more dynamic aspects of television. This person will need to understand the craft of the performer and director in order to design settings that will enhance the action taking place within them, help the audience to accept the performance choices made by the actors and director, and make useful contributions to the planning phase in which such choices are made.

Typically, students of art or interior design learn to work in static media—studying composition, color, form, and so on—and are often better able to produce clean and very salable renderings of design proposals. There are several traps into which they may fall. One problem is lack of experience with camera and performer dynamics and a tendency to produce settings that overwhelm the performance rather than support it. The need to study (and master) the subtleties of camera composition and dynamics—together with the shortcomings—is, of course, something that students of art or interior design have in common with those who come to TV from a theatrical design background. The other pitfall stems from the idea that the work of the designer is the centerpiece of the production, rather than a subordinate part of it. This, together with the idea that the design rendering is itself the finished product—not merely the initial step on the way to a more complex goal—are the most common ways in which graduates of art courses tend to trip up.

These are the two extremes of training from which most television production designers come. Television design represents a compromise. On the one hand, those who come to it from theatre tend to have a stronger understanding of the dynamics of the medium and a stronger technical background (a better understanding of the various support services that go into the mosaic of production), but a tendency to dismiss the skills of drawing and rendering as unimportant. On the other hand, those who come from an art or design background are usually better at illustrating their concepts and less interested in (and often less patient with) the preproduction planning and the construction phases. Nonetheless, an ability to think visually, to express oneself clearly in visual media, and a willingness to study the process of TV production are the only essential qualities to successful pursuit of a career in production design.

Training

It is unfortunate that few colleges are teaching production design for TV; many are still training students for nonexistent jobs in theatre design, and many students studying design in art departments are doing so because there are no courses available in design for television or film. Hopefully, things are changing and academia will soon catch up; however, those who are enrolled in other types of design courses with an eye toward making a career in TV can be encouraged by the fact that most of those working in the field today had to make one of those adjustments I've discussed. Finally, there is no college course (no matter how well taught or by whom) that can substitute for on-the-job experience. The most important aspect of job acquisition in this field is the portfolio: Many of those who have succeeded have done so without any college education at all. The traditional way to gain training in this business was through apprenticeship to an established master; this sort of one-on-one education is still the most effective, and anyone (including those with advanced degrees) would do well to begin a career in this way.

New Directions

The major television-producing countries are trying to negotiate international standards relating to high-definition video, but those closely involved with this end of the medium are not optimistic about the prospects for success. Politics, protectionist industrial lobbying, and the desire to make a new system compatible with existing broadcast formats (a problem verging on the impossible) all mitigate against the establishment of a video format with the widespread acceptance of film. As in the past, those working with production have adopted the new medium in a limited way and have adapted it to suit their needs. Some commercial and a limited amount of feature production is being originated on Sony 1125 instead of film. It is used only as an origination format, with the intent that it will be down-converted to broadcast standard for transmission after the project is edited. The advantage of the Sony 1125 is that image quality remains high throughout the production and editing stages, and thus delivers a better quality picture at the showing stage. This has several tangible advantages over film: immediate playback at the time of shooting; simpler electronic editing; simpler (and much cheaper) special effects; and a much reduced cost over film in terms of material, studio, and labor.

It is not inconceivable that such a videotape medium might replace photographic film in cinemas, because it can be projected quite successfully and provides resolution as good as film. Until that time, though, it is possible to do a laser transfer resulting in a film print from an 1125 video original. Thus it may be possible to substitute this format for film in both feature film and film-for-TV contexts, and both of these uses are beginning to be accepted in the industry. It is also conceivable that feature films could be distributed from the production company's studios to the local cinemas or to regional distribution points via satellite, greatly reducing delay and cost.

Some people are referring now to high-density television, the difference being that this term refers not only to increased capacity for detail in the picture but also to increased capacity to carry information on sidebands and the resultant possibilities for viewers to interact with their television programs in a direct and stimulating way. The options here are limited only by the imagination. Whatever the future brings (and past experience should tell us the futility of attempting to predict such things), it will likely be technical advances such as those I've described. What is likely to remain unchanged is the body of things that motivate us to perform, that motivate us to tell stories, and that make participation in this process enjoyable for both the artists and the audience. Special effects will become easier and less expensive to produce, and they will be increasingly more convincing; but the pleasure derived from being fooled by them, the excitement of the fantasies they bring to life, which has been with us since the outdoor drama of the ancient Greeks (or longer), will not likely change or diminish as long as humans continue to exist.

While it is essential that the artists working in television stay abreast of the technical developments, it is equally important that they understand that these things are tools to assist in the achievement of a thing—a fleeting moment of drama, excitement, laughter—that we have pursued with whatever tools were available for thousands of years. The best and most successful artist is the one who knows the difference between the tool and the objective, the one who is familiar with the work of fellow practitioners from the past, and the one who understands not only the form within which he or she must work but also the process by which it evolved.

Roles and Responsibilities 3

Staff

In the old days of television and film, these media had their own terminologies when referring to job titles and equipment. Since crossover between them and the theatrical crafts became fairly common, however, the terms of reference have begun to change. For years, it was the norm in television to refer to virtually everyone working in the studio as some sort of director. In addition to the actual director, there was the technical director (who operated the switcher), the floor director (floor manager), the lighting director, and the art director. These terms never caught on in the European system, which refers to the technical director as the vision mixer, the lighting director as the lighting designer, and the art director as the production designer. The importance of titles tends to escape those who are not directly involved, and may seem like a petty ego-gratification, but there is a practical distinction to be recognized.

In traditional television, the star system has been as strong as (or stronger than) in film. While the great film stars of the 1930s were household names, they were virtually owned by the big studios, being held to strict contracts for long periods of time. When TV came along, the powerful studios were a thing of the past, and television—with its tendency toward quick ascents and equally quick declines of stardom—was not in the business of managing its creative talent in the way that the film studios had done. As a result, a star such as Jackie Gleason or a strong producer such as Fred Friendly would have a great deal of content control over a given program. In such instances, the other production personnel, from the director on down, would tend to become functionaries working for that one strong person rather than creative partners. The roles of these other production personnel were regarded as generic ones in most instances, and the individuals assigned to them might change from one week to the next without anyone expressing concern for artistic continuity.

Thus it reflects a revolution of sorts within the industry when we see credits for production design and lighting design rather than art direction and lighting direction. This is an indication that the roles of these people are expanding and that the industry is acknowledging their increased contribution to the creative effort of the production. It also indicates that the scale of production is changing and that the importance of visual subtlety is growing; increasingly, the audience for television is less content with hastily prepared superficial comedies, and the talents of those who can produce high-quality visuals for more complex productions are now in demand. So it is that the production designer is replacing the art director.

Production Design

Typically, a production designer is responsible for the supervision or execution of most of the aspects of program visuals. In past practice, the art director would have been a set designer, with set design being the main area of concern. The production designer is responsible not only for the settings, but for the costumes, props, makeup, and sometimes lighting and graphics. It is not entirely necessary to be expert in each of these fields, but one must be able to converse well with each department and to command the respect of the craftspeople involved. The production designer's responsibility is for the entire look of the program. In order to work successfully the production designer must be capable of originating a style which will identify a particular program in the public mind, from the look of the lighting, sets, and costumes to the typefaces used in the opening graphics and the closing credits. To do this, it is necessary to stay in touch with evolving graphic-design artistry and technology, fashion trends, furnishing styles, architecture, and sometimes—as in the case of science-fiction productions—the fashions of cultures that don't exist yet.

As the scale of production grows, the size of the support team also grows, and the production designer becomes the manager of a rather large department as well as a member of the creative staff. Working immediately with the production designer is a staff of assistants, draftspeople, and model-makers who help with the rendering of the design ideas and coordination of the other support departments that depend on them for the information required to do their work. These people may be variously referred to as design assistants, set designers, art directors, or model makers depending on the locale and the specific assignments.

At the high end of production, all of the crafts and trades that provide production services are unionized and strictly defined. Membership in the appropriate union is required in order to work, although many unions recognize the value of apprenticeship and encourage entry-level people to work in various assistant jobs in order to learn the craft and make their way up through the ranks.

One of the first support services contracted is the scenery construction shop. In many cases, this is an outside contractor who does nothing but provide settings for studios. Such companies are usually built around a core staff of carpenters and painters, and perhaps a few metalworkers. The bulk of the work they do is wood construction, making flats and platforms and finishing them to the specifications of the designers. Scenic carpentry is very different from carpentry used in house construction: Durability is less important than portability, so some structural strength is sacrificed for lightness. A TV flat is typically built from a sheet of ¼" plywood—the facing—framed on the back with 1" × 3" pine. These will be used to represent walls, and will appear quite solid (and withstand fairly hard impact when properly braced) and yet each panel can be easily moved by one or two stagehands.

The scenic artists are members of a craft that traces its roots back to Renaissance Europe, and the skills required have changed only slightly through the years. While the surfaces upon which they work have shrunk very much from the huge canvases that were typical of opera in the nineteenth and early twentieth centuries (and are still seen in the great opera houses of Germany and Italy), the ability to accurately render a style of painting to full size from one-half-inch scale renderings is still the essential skill. While it was often necessary to reproduce effects such as that of wallpaper with paint and stencils to make them seem real over theatrical distances, it is now possible to use the real thing in such media as film or television. The more subtle skills, such as painting realistic woodgrains and accurately rendering the effects of weather or age on various surfaces are still highly prized and very much in demand.

What makes a studio set believable to the audience is the way it is dressed; how well the material details of everyday living are represented. This is also the purview of the production designer, but the realization of set dressing falls to the property department. Props people tend to be divided between those who work in the studio cataloging props and putting them in their proper places on the set and the buyers who are responsible for locating and acquiring props. There is room for a lot of creativity in the prop buyer's job: A familiarity with period styles (meaning anything other than that which is currently in vogue) and a resourceful nature will quickly make a prop buyer a good reputation. It is immensely helpful to the design team to have a buyer who can get the job done well with a minimum of supervision; one who has a good eye for detail and a knowledge of style will prove invaluable. In the larger cities, and in areas that have been home to theatre, film, or TV production for some time, there are established prop rental houses that often have huge inventories of the most disparate and obscure artifacts from every period imaginable. In some cases, such as Robert Altman's film *Come Back to the Five and Dime, Jimmy Dean*, the precise accuracy of period props becomes critical, and prop construction specialists are called in to fabricate all sorts of odds and ends from everyday existence of another time with flawless accuracy. This, again, is based on the research of the design staff, although there is plenty of room for advice and input from the props artist, and here, too, informed creativity is always an important asset. Once the props are acquired, they are assembled by the prop master, who is responsible for cataloging and organizing every item. This extends to the identification of the settings on which each prop is used and the placement of it thereon. Although most productions have a script supervisor to help with continuity, the final responsibility for continuity of set dressings from one shooting session to the next falls to the props and design crews. Typically, this is achieved through liberal use of Polaroid film and a large photo album. Props are further divided into set props and hand props, the latter being those that are carried into shot by an actor. Hand props are kept on a table off-set, and handed to the performer as needed and then returned promptly to the prop store after use.

A note about continuity: It would be easy to shoot television (from an organizational standpoint) if all the scenes were shot in chronological sequence. This is rarely cost effective, and sometimes simply impossible. Scenes are usually shot one location at a time, and out of sequence, since that is most economical. A particular set will be erected in the studio, dressed and re-dressed several times as all the scenes that take place in that location are shot. When that is done, another set will be erected, dressed, and lit to shoot another sequence of scenes that take place in that locale. This method is highly efficient and makes the best use of very expensive studio and crew time. It requires that the director's staff and the design staff make rigorous and highly detailed notes about the dressing of each scene, and keep absolutely accurate records of each scene as it is shot for future reference. Each department is responsible for its own continuity, seeing to it that details of lighting, wardrobe, sets, props, and makeup are consistent throughout the production. This is, obviously, most critical in scripted dramas, but is also important for news, talk, game, and variety shows; consistency of product is something the medium has strived to deliver, and its audience has grown to expect.

Costume Design

The costume designer is another production design team member who has grown in prominence in recent years. Sometimes the production designer is also a fully qualified costume designer and he or she only needs a wardrobe staff to properly service a production's needs. For news programs, for example, the designer selects wardrobe items off the rack in clothing shops for the on-air talent and these items are held in the station's wardrobe stock and maintained and cleaned by the wardrobe staff. In some smaller stations, the talent will be given a clothing allowance and required to keep and care for their stock of clothing.

For scripted drama or comedy production, the costume designer takes on a much larger role. Though nominally responsible to the production designer, the costume designer is a specialist and generally in charge of a costume department. Costumes must be more finely detailed for television than for theatre and must stand up to much closer scrutiny. Over the years, clothing styles have undergone hundreds of changes, which are reflected in every aspect of design from silhouette to different ways of sewing seams together. The costume designer must be familiar with all of the historic styles of dress in detail, including such costume accessories as jewelry, handkerchiefs, purses, eyeglasses, and so forth. The costume designer also needs to understand the cutter's patterns of each style and be able to draft patterns for the costume construction crew to follow in cutting and assembling each piece. The costume designer must know all types of fabrics available and know (or be able to invent) ways to simulate the textures of fabrics that are no longer manufactured. In addition, this person must draw accurate sketches of each costume (complete with samples of all the fabrics intended for each) and construct a costume plot showing what costumes are needed for each character in each scene and on which day of the shooting schedule. Also, because many production budgets do not include hair or makeup designers, the costume designer often assumes these roles.

Most costume design courses are offered within the curricula of theatre departments. The differences between theatre and camera work are somewhat less substantial for the costume designer than for the set designer, and people with training in theatrical costume design have long been crossing over into film and television. As with set design, costume design requires an extensive study of the history of style. The costume designer needs to understand the evolution of fashion, from shape and pattern to shoes, combs, and textiles. It is critical to know about the life-styles of a given period—to understand how people walked, traveled, sat, danced, slept, and swam—in order to make the costumes look authentic and move correctly. The costume designer and the wardrobe staff work closely with the director and the actors, instructing them on the correct wearing and use of costume pieces.

The wardrobe staff can include a number of skills depending on the complexity of the production. There is always a wardrobe master or mistress whose job parallels that of the property master: cataloging, storing, and distributing the various costume pieces, and keeping records for continuity. There is always a repair and construction crew, skilled and knowledgeable in sewing and trimming techniques, who build any costumes that must be constructed from scratch and repair any costumes damaged in use. If all or many costumes must be built (the term is *built*, never *sewn*) for a production, a separate crew may be hired to do this, and another (smaller) crew kept on for repair and maintenance during production. These may be employed by a studio or by an independent costume company. In either case, there is usually a large stock of costumes from previous productions from which pieces can be pulled and modified to suit a new production. Most costumes for a production set in the present or the recent past can be pulled from the racks of retail clothing stores. If the program offers sufficiently broad exposure, clothing manufacturers can sometimes be persuaded to donate wardrobe items free of charge (usually for an on-air mention in the closing credits). The larger costume houses and construction shops in the larger studios typically employ cutters, whose specialty is the accurate cutting of costume parts from scale patterns or even (if they're very skilled) from a costume designer's sketch.

The costume designer works very closely with the director and the production designer, taking an important role in defining the look of the production. Whether it is a weekly magazine show or a full-blown TV drama, the costumes or clothing worn by the talent is a major compositional element because it is nearly always sharply in focus and shot close up. Clothing helps to define the character of a given person on camera and is a primary factor in the visual process by which the audience decides how it is going to react to a specific person. An anchorwoman reading the news would typically not wear a frilly flower-print dress with a lace collar any more than she would wear a slinky sequined evening gown; neither of these encourages the audience to view her as an authoritative figure and thus to take what she says with the seriousness appropriate to the evening news. Similarly, an actor in a drama or a comedy must be dressed appropriately to the character, and the entire production must take on a costume style and texture that is right for the story.

Lighting Design

Both the settings and the costumes are at the mercy of the third major element of design—the lighting. In the past, each production had a lighting director whose task was usually defined in terms of illumination and who was regarded as a member of the technical staff. Certainly, this was adequate in some situations, especially when TV was shot in black and white, and cameras were not very sensitive. As cameras became more sensitive and color became standard, audiences began to expect more subtlety from the lighting. As a result, the lighting designer has come into the studio and come to be regarded as a creative contributor rather than simply a technician. In many cases, these were lighting directors who took the time to learn more about the art of lighting, and thus combined technical expertise with artistic technique. Others came from theatre and film backgrounds.

Lighting is some years ahead of the other design areas because it has been taught as a component of film and TV courses for a long while. It is still possible to learn creative skills in related fields and transfer them to lighting; and I know of some highly successful lighting designers who studied painting and sculpture. At its best, lighting is a sculptural art: It is a challenge to make things portrayed in two dimensions appear three-dimensional, a challenge of sculptural technique. It is also a challenge of color perception to light colors on a set, colors on costumes, and colors of complexion to achieve just the right mood and atmosphere. Light is the most intangible medium used in production design and also the most vibrant and dynamic; to many, it is for these reasons both the most challenging and the most rewarding of the design areas.

The lighting crew is usually small and hardworking. While the other departments have weeks or months to build sets and costumes, the lighting crew has hours (or, if it is lucky, a day or two). The lighting process is divided into three phases: planning, rigging, and focusing. After the set has been designed and the costumes drawn, the lighting designer will meet with the other creative staff to plan the placement and function of the lighting equipment. They will discuss colors, placement of talent on the set, and desired effects while the lighting designer works out possible solutions in her head and tries to anticipate problems (at the same time making notes for later reference).

The lighting designer will then go away and prepare a plot, based on the production designer's floor plans, indicating the exact placement, function, and equipment type for each light source. When the plot is done, an inventory of all equipment by type will be prepared and compared with the inventory of equipment available in the studio. Those items not already in the studio must be ordered from a lighting rental company for delivery on the rig day. Studios sometimes come complete with a large lighting inventory included in the rental, but studios may also be rented completely bare. Location shoots, obviously, must be completely equipped, literally from the ground up.

On load-in day, the equipment required for the production is installed in the studio. Sets are erected; props delivered; and lighting equipment brought in, tested, and inventoried. When the set is in place, the electricians begin hanging all the equipment in its appropriate spots and connecting it into the studio's electrical system. Depending on its age and the original use for which it was intended, the studio may or may not include a dimming system.

The electricians are usually responsible for all power points on the set; if there is a practical lamp, a TV set, or a blender required they will wire it up. For items such as this, they will need to work closely with the production designer and the props crew, as well as the director and the floor manager.

In the strictly unionized studio, demarcation of crew responsibility is absolute. A props person may place a lamp on a table, but an electrician must wire it, turn it on, and disconnect it after the shoot. Props people, likewise, do not handle costumes or permit any other crew members to handle the props. In the nonunion or partially unionized studio, job areas are not strictly defined, and a floor manager may find himself doing props, lighting, and some wardrobe tasks or even operating a camera at times.

Television production is a labor-intensive business; for every head appearing in front of the camera, there are many more behind it, whose work began weeks or months before and will continue long after the performers have gone home. The designers and their teams are a major part of this component and although the size of the design staff varies greatly depending on scale and budget, their contributions are essential and visible at all levels. Often, the value of a good design staff is noted by its absence, although the trend is now toward more attention to design considerations and more budget support in this area. As the audience becomes more sophisticated and the medium more competitive, production design, lighting, and graphics become important tools in the competition for the critical audience share—and rightly so. From the high-end drama series to local news programming, the achievement of a look that supports and helps to define a program style is being given more attention by the producers of television programs, and the skills of people trained in these crafts are in greater demand.

4 Objectives

The Bottom Line

In the context of commercial television, which is the predominant form the medium takes in the United States, there is really only one objective: profit. The industry earns its profits from the sale of advertising time. The price for a given unit of air time is based on the estimated size of the viewing audience, as determined by one of the rating services. In a somewhat roundabout way, and with some large gaps in the system's means of assuring accuracy, the equation is this: viewers equal profit.

The ratings wars are cutthroat at virtually every level of the market, and every possible tool is drawn into the process. Even though most local stations buy the majority of their programming from other sources (either networks or syndication companies), they still use any means available to attract more viewers than their competitors. In many cases, the only significant home-produced program is the news, and it is in this arena that the pitched battle is fought.

The local news show is often operated as a loss leader, costing much more in resources and salaries than it generates in advertising revenues. This is because it is often the only significant locally produced program and, therefore, the only arena in which local stations can compete directly. There is also a spillover effect; the evening news is usually the first program in a typical household's evening of television, and that audience may tend to stay tuned to a given station for the remainder of the evening if they had tuned it in for the news in the early part of the evening.

The Commercial Role of Design

Part of the competition among stations is the creation of a station identity, an image. The overall profile of its viewers will be taken into account: The station will hire newscasters who fit into the age and life-style image of that group, and will use its other image-creation tools to encourage viewer identification and loyalty. All of the elements of design are enlisted in this campaign: the color choices, the graphic styles, and the overall look of the program (including the dress of the on-air talent).

Design plays a very important role. Even in a regional station (though not, usually, in a small-market one) it is not uncommon to find a full-time production designer whose principal responsibility is the news program. In the case of an O and O (a station *owned and operated* by the network), there is often a network look developed at the national level, and adapted for each station's local news program.

The industry knows the importance of the designer to the process of image creation and the importance of image creation to the task of selling its programming to the viewers. The best example of this was the creation of the Fox Network in the minds of the television public. Having identified a younger, less conservative audience as being neglected by the major networks, Fox proceeded to build an image for itself of a youthful, upbeat, and trendy entity in opposition to the stodgy and predictable majors. Fox took some risks with program choices, such as the self-deprecating humor of Garry Shandling's show, and the revival of the prime-time cartoon show. The film-noir look of shows like "Booker" and the blatant cynicism and sexuality of "Married With Children" were designed to appeal to the tastes of the under-40 viewers, and their success proved that such a disenfranchised audience did exist and was hungry for media recognition.

Both "Garry Shandling" and "Married With Children" used the tried-and-true look of the classic sitcom, though in both cases it was a self-conscious choice intended to work as a spoof of that form. "Booker" took off from the point at which its predecessors, such as "Miami Vice" and "Hill Street Blues" had begun. "Hill Street" had been innovative in its time through its refusal to dress up reality to make it more palatable to the home audience. Scripts rarely had neat conclusions; stories were left unresolved, and, often, the good guys lost the fight or were found to be less than good. All of this was reflected in the look of the show: The locales tended to be back alleys and dilapidated buildings and the interior sets had the look of old and dirty public spaces. A great deal of this look was achieved by attention to detail; there was always a visual subplot happening in the background of the shot, which added to the realism of the scene.

"Miami Vice" picked up on these possibilities, combining the seedy with the flashy and high tech, a reasonable approximation of the world of high-rolling drug dealers. The color palette was very carefully constructed to feel like Miami, and the look of that show was so popular that(for awhile) men everywhere were trying to get the "Don Johnson look."

Design and Advertising

Thanks partly to television, we are a visual culture, and much of our self-image is defined in visual terms. This is good for business; Look-conscious people will spend more on new clothing, new cars, new places to socialize, and so on. The TV set is the first place we are approached with trendy new things to buy and is the arena in which much of this battle is fought. So it is that those designers working in commercials are among the most sought after and best paid. The ability to create an immediately recognizable look, and one with which a target audience will identify and want to emulate, is critical to the success of advertising, especially with products that are luxury items rather than necessities. Cars, perfumes, clothes, cameras, and such are nearly always sold by means of identification with a life-style, and are thus substantially dependent on the ability of a designer to build the aura of such a life-style around the product.

Among the most competitive television advertising arenas is that of the car commercials, and it's a good place to do some armchair analysis and draw some conclusions. The pitch of Volvo commercials has long been reliability and safety; the same has been true of Mercedes-Benz more recently. In both cases, the atmosphere of the ads is always rather high-key lighting, generally somewhat flat, such as might be found in a scientific research lab. These commercials often show real test lab footage of cars being crashed into concrete barriers or dropped from extreme heights, presumably to test structural strength. They often feature actors dressed in lab coats and carrying clipboards on which they make copious and (perhaps) meaningful notations. In these scenes, the surfaces in shot tend to be smooth and solid-color, and the highlights are usually chrome (or some other metallic surface) or such things as scientifically significant yellow lines on the floor. All of this is intended to create an aura of high-tech mystique and authority, to lend credence to the manufacturers' claims to superior safety standards. While it is highly likely that these claims maybe true, our culture no longer finds it adequate in its advertising simply to state the facts and rely upon the viewers' good judgment. In a way, this is good for the designers: In the old days, advertising was frequently no more visually sophisticated than a talking head standing in front of a drape, and no design services were required.

We've come a long way since then. At the top end of the industry, huge sums of money are invested in the creative aspects of 30-to-60 second

Figure 4.1 A commercial for Honda: "Ya just can't make this stuff up." Courtesy of Ron Weiss, Korey Kay & Partners

bits of airtime, and much of the budget and responsibility falls to the design staff. Let's look at a few other commercials in the car sales business. Some show a fast car negotiating country roads while classical music plays, others show them in fantasy racing scenes, and still others don't show the cars at all—just someone talking about the feeling of owning one. In all cases, the design elements have been carefully chosen to build the correct aura. The country road scenes tend to feature softer color contrasts, sometimes autumn leaves, and use the full range of sensitivity of film. The hot-car commercials tend to use a brighter, more contrasty color palette and more extreme camera angles together with stronger and more angular lighting. The third type, such as the Honda commercials produced for the New York regional market, may use a single shot of a realistically portrayed person in very realistic surroundings simply talking about the car. This last example used design most extensively and was (to my mind) more effective. It featured a New York cop in a set that was a very realistically dressed and detailed representation of a city police station (inspired, no doubt, by "Hill Street") and consisted of a slow zoom in to the cop's face while he described the enigmatic case of a car thief who only took Honda Preludes, and always replaced them with older models of the same car, finishing with the line, "Ya just can't make this stuff up." The psychology of aura creation was superb: The viewers were given an environment with which they were familiar and with which they were able to associate, and a typical city policeman, who also fit the pattern of familiarity. These elements combined with the realism of the scene to make the message being delivered by the actor playing the cop both accessible (familiar) and authoritative (the uniform and the context of the scene) and quite likely more effective in building the car's image than anything they could have shot that might have shown the car.

The same agency also produced a series of commercials for the same client that only showed other manufacturers' products: In each case, a very detailed and realistically portrayed scene showing a sweating driver pushing a broken-down car, or with the hood up and the sleeves of his expensive dress shirt smeared with grease. In all of these cases, the ability of the design team and the photographer to create images that conveyed exactly the right atmosphere set up the framework within which the viewer was primed to accept the text of the commercial as factual and believable.

To return to the theatrical comparison, the objective again is the suspension of the audience's disbelief; in order for the theatre audience to become involved in the action of the play, and in order for the television audience to accept the words of the person pitching a product, the audience must be able to accept the context of the presentation as real and believable, and the skills of the production designer act on both in the same ways.

Clarity in Objectives

It would be unfair, and certainly less than a complete analysis, if we left our discussion of motivation at this point. The driving force behind all of commercial television is commerce; nonetheless, we should admit that art plays a major role in this industry and one of the chief objectives of the creative people producing television programs is the achievement of some sort of aesthetically satisfying work. It is often true that the design staff will see themselves as working in opposition to the front office, whose overriding objective is cost control. (To be fair, those in the position of financial control tend to see designers as profligate spenders.) Although the management duties of the production designer frequently take over the time one feels ought to be devoted to creative work, the target goal typically uppermost in the designer's mind is the creation of a cohesive visual package that supports the substance of the piece as much as is possible within the constraints of time, talent, and budget. There is little time available to devote to the achievement of this objective, and it is a triumph of determination when the designer succeeds; there is almost never any time left over to worry about the salability of the show or its ratings. Presumably, one of the contract obligations of the producer is the growth of ratings-related ulcers, thus freeing the creative staff to grow their own, worrying about construction deadlines and last-minute changes of staging requirements.

While the producer may have such esoteric worries in mind from the outset, it is her job to assemble the creative talent at the initial production meetings and help them define the objectives of the project in the even more esoteric terminology of art. This is true from that point forward, whether the project is being produced for a commercial entity or a noncommercial one. Directors like to think in terms of telling stories and designers in terms of composition. The common ground is the fact that everyone involved is building something and all are aware that it will only hold together if they all are building the same thing. The determination of the nature of that thing is the responsibility of the director and often the producer, too. The process of identification of the common goal and the whole process of realization will go much more smoothly if these two people are visually literate and able to describe their objective in visual terms however vague they may be. It is not possible, however, to rely on this—the rare occurrence of such a dialogue will teach you to treasure those with whom you achieve it and encourage you to work with them again.

Having identified an objective in somewhat visually stimulating terms, such as the colors of the sky at certain times or during certain seasons, or by relating to a piece of music or a painting, the group will go in separate directions and research the ideas that seemed most appealing in the first meeting. They will then reconvene to compare notes and findings, look at resource materials and rough sketches, and so forth, comparing reactions and further refining the objectives of the project in visual, emotive, and narrative terms. This is a cyclical process, which will be repeated many times until the outcome is agreed upon and thoroughly understood or until the deadline for the commencement of production is reached. In most cases, whether for commercial or noncommercial production, it's the latter reason. Left on their own and without an accountant looking over their shoulders, most creatives will produce a flawlessly detailed piece that unerringly portrays the universe as seen through the eye of the scriptwriter, the story impeccably told, but several million dollars over budget and running four hours long. I hate to admit it, but my objectives in a design job have never been so sharp as when an externally imposed deadline loomed on the day-after-tomorrow horizon; it is a great motivator and stimulant to clarity of thought.

The objectives defined in preproduction meetings will reveal an evolutionary process for the designer; it may be a mutually cooperative narrowing of focus or it may be a tug-of-war, but you will emerge from it with an agreed-upon goal as far as the look of the piece, and the amount and type of work you will be required to do to realize your objectives. It is essential that agreement be reached; while your work is artistic, it is also a critical component of a business arrangement and you absolutely must be in agreement with your director and producer as to the desired outcome. While it is usually best to drop out of a project rather than attempt to press on in a situation in which you find yourself irreconcilably at odds with the director or producer, it is also important to the preservation of a good working relationship to see to it that your agreed-upon criteria are clearly laid out and, whenever possible, laid out on paper.

Commercials, because they need to accomplish so much in so little time, make an easy example of a situation with a clearly defined objective. The high-budget network shows also fall into such a category; much more difficult is the situation in which much of the show's content is unknown or poorly defined. Talk shows, news shows, and variety shows tend to cover such a large scope that defining the objectives is hard to do beyond finding a look and trying to keep up with the visual trends. In such cases, the only peg on which to hang your ideas is the functional format of the show. This is, in its way, an objective: the object in such a situation is to create an environment that is friendly to the desired format and adaptable to those variations and possibilities that can be anticipated. Sometimes, the objective is met by building in as much flexibility as possible. A talk show, for example, may anticipate that the usual format will be a one plus one (an interviewer plus a guest), but may occasionally expand to become a one plus two or a one plus three. The classic "Tonight Show" setup does this; there is always one host and a variable number of guests, hence the desk/chair and long sofa combination. This permits limited interplay among any number of people, but favors the one-on-one interview at the camera right end of the set. The look of the set tends to be casual and the lighting high key, both of which are conducive to a relaxed and informative discussion. There is also a performance area in which musical guests or comedians can do their acts. In this area, valiant efforts are made to design a set that is supportive of the image of a given performer; this is rarely very successful, particularly when held up against the production values of most music videos. A significant negative example was the short-lived and ill-fated talk show hosted by Pat Sajak. Just prior to the show being canceled, the set was redesigned and refurnished, (in my opinion, inappropriately), indicating a lack of a clearly defined objective: Even in the superficial terms of talk and light entertainment, the lack of clear image objectives can be fatal.

When NBC instituted "Saturday Night Live," it made the objective clear: The live element was to be adhered to almost 100%, and the show would play up the tension and immediacy inherent in any live performance. As a result, they built a core of repertory players and comics/writers along the lines of vaudeville, building in a lot of music and topical humor. The design of the studio set was reminiscent of a theatre, and shots were often used wide to show the performer–audience relationship and reinforce the theatrical context in the minds of the television viewers. It was an indication of the clarity of the show's objectives that the producers hired Eugene Lee, who had extensive experience designing on and off Broadway, and was well known for the strong theatrical look of their work. Though the sets have undergone renovations over the years of "SNL's" history, the identifiable look of the show has been carefully retained, and Lee is still the production designer. The objective of the design was to find a look that would evoke the atmosphere of a theatrical context with which the audience was already familiar and within which the sort of comedy-variety-music mix would be acceptable and appeal to the late-night crowd.

The ultimate objective of any designer is the old one of suspension of the audience's disbelief. If you can get the audience to accept the terms of your presentation, they will be drawn into it and they might not change channels. Whether the program is being broadcast by a commercial or a noncommercial entity, or whether its purpose is to entertain, inform, or educate, the desired effect can only be achieved if someone is watching. Jon Stone often defended "Sesame Street" against critics from the educational establishment with the argument that entertainment was at least as important as education to the program—they could only educate those who were watching, so the first priority was to keep them watching. One of the most useful tools in this is design; it can be used to target a program by cueing the viewer as to the type of show she is watching, and it is necessary to keep the viewer by setting the proper context for the events depicted in the program and making them believable. Badly designed and shabbily presented programs may not cause viewers to turnoff their sets, but may well cause them to tune out; when you've lost their attention, you've failed, and the production designer's role in this transaction can be very important and constructive. Critical to the success of the design is a clear articulation of the program's objectives and agreement among the creatives as to the best means to achieve them.

Composition 5

Controlling the Focus

You will recall that the theatre set designer is faced with the challenge of creating an environment with which to direct the focus of the audience. In an arena-style production, this is somewhat more difficult than with a proscenium; across from any given audience member is a group of other audience members, any one of which might prove to be more interesting than the action of the play. In a proscenium theatre, the audience members are all facing the same way and so are more inclined to pay closer attention to the action onstage. Additionally, the proscenium opening provides a frame within which to construct the design elements. This all adds up to control; the window through which the audience views the action narrows their focus from everything in the theatre to only those things contained within the proscenium, and greatly eases the task of controlling their attention.

Proscenium openings have no standard size or shape; they can be quite high and narrow (the operatic tradition) or they can be wide and low (the Cinemascope tradition).It is often the case that a play (such as Shakespeare) may demand a higher, narrower format, but the designer must find a way to fit it into a long, shallow frame. What choices must this designer make in order to adapt? The excess space at the sides may be filled in with some nondescript masking and the set brought forward of the proscenium opening, with a balcony built over a temporary thrust stage. This normally results in fights with the producers over revenue lost due to the reduced seating; thankfully, a challenge that the TV production designer rarely (if ever) faces.

Whether it's 14 inches across, or a 20-foot wide video wall, the television screen is always three units high by four wide. This is called the aspect ratio, and it doesn't vary—NTSC, PAL, and SECAM are all 3 by 4. It is yet another factor in modern television that was locked in during the medium's infancy, in the 1930s. Conventional wisdom says that this was an unfortunate choice: While the format is great for a head-and-shoulder shot of one person talking directly to the camera (hence the term *talking head*) it is a very difficult format for wide shots and especially for shots with two people in the frame.

There was disagreement between Steven Spielberg and the issuer of the video version of *Indiana Jones and the Last Crusade* over the cutting of a two-handed scene between Harrison Ford and Sean Connery. This scene

Figure 5.1 TV aspect ratio

had been conceived for film and shot as a single continuous take; in order to convert to television format without letterboxing it was necessary to convert the scene to a series of left-facing then right-facing profile shots, disrupting the original compositional choice and changing the aesthetic tenor of the scene. This is only a recent example of a problem that has plagued those people in the home-video movie market since its inception.)

The film industry has experimented with different formats for many years. In the 1950s, there was a short-lived phenomenon called *Cinerama*, which used multiple cameras to achieve a surround effect. It was crude (you could see the seams between the shots) and it was quickly proven to be a passing fad, but it pointed to the desire of filmmakers to include everything in shot—to their frustration with the limited frame within which they were restricted. This is a subject that has been discussed frequently in the industry and has been sufficiently important to warrant expensive camera and projection system developments that were then discarded. We may take it as significant, then, that when given the opportunity to redesign the television format, the people developing HDTV chose to make the aspect ratio 5 by 9, very nearly identical to that of anamorphic film. Nonetheless, since the adoption of HDTV as a broadcast format looks unlikely, anyone entering the field of television production design now can expect to deal with the old and cumbersome 3 by 4 ratio for some years to come.

In its simplest application (and probably the one for which its originators intended it)—the basic talking head of the newscaster—television works pretty well. It is a shot so common that there are hard-and-fast rules of shot composition about which few will argue. Typically, this shot includes the head and upper torso, with the line across the eyes (*eyeline* is a term you will encounter very often) cutting across the frame about one-third of the way down from the top. This, together with its modification—the talking head with graphic insert over the shoulder—are the staple shots of the news program.

Talking Heads

The news program is the situation in which the design of the setting is most prominent to the eye of the viewer. It often includes some sort of station identity material such as a logo, graphic representation of a city skyline, or other easily identifiable symbol that must be clearly seen and recognized to function properly. The trick in this case is to achieve a balance between recognition and subordination of focus within the shot; the station or program logo on the backing flat must be easily identifiable, but not so strong visually that it steals focus from the talking head in front of it.

Because this format is the most strictly defined, it is also the simplest to use. If the shot is a newscaster simply filling the frame and reading to the camera, the composition consists of the head centered in the frame with the eyeline about one-third of the way down. It is not necessary to

---

Letterboxing is the addition of black margins to the top and bottom of the screen to preserve the film aspect ratio.

Figure 5.2 A talking head

know what the talent is holding in his hands; we want to see the face, especially the lips, clearly so that the enunciation and inflection of the voice can be most strongly reinforced. (Jane Curtin made an excellent comic use of this situation some years ago on "Saturday Night Live," describing in graphic detail, and without changing from her newscaster's monotone, what she was wearing underneath the desk where no one could see.) The production designer's greatest concern in this situation is one of scale and texture. As noted, the treatment of the news backing must be recognizable in shot, but not obtrusive. If it is a station or program logo, it needs to be contrasted from the base tone of the background enough to be readable, but not so much as to come across in shot as a higher contrast image or a brighter image than the face of the newscaster. Size is also an important consideration; if the lettering of the program logo is greater in height than the face of the in-shot head, it will likely command greater attention and thus fail to work. A common solution is to make the logo into a pattern on the backing, painted in a tone that is fairly low contrast in relation to the base color and yet still strong enough to be clearly readable in shot.

Graphics

Now let's raise the ante and insert that classic over-shoulder graphic that is so frequently used in news programming these days. Since it is generated through the video system, it is also usually a three-by-four rectangle. When used best, it fits into the upper corner of the screen with a reasonable margin between it and the head on one side, and between it and the top and side of the frame on those sides. The margin between the graphic box and the head is most critical, because the box is often electronically overlaid on the studio shot and will cover some of the presenter's hair or face if too little space is allowed. (How many have never seen this mistake on the air?) The margins on the other two sides are also important: If the box is

---

Your design succeeds or fails in the context of the shots which were (or should have been) decided upon in preproduction. No one cares if the set looks good from the studio floor, because no on except for you and the floor crew will ever see it from that vantage point. Always assess your work from the camera's point of view and *only* from the camera's point of view.

Figure 5.3 Headroom and margins

touching the side or top of the picture it looks wrong. (We are accustomed to seeing pictures in frames, and some framing space makes the image within it stronger.)

In such a shot, the camera must zoom back somewhat to make the head smaller in the frame and allow room for the insert. This will necessarily include more of the torso and probably the hands and desktop in the shot and reduce the size of the background treatment on the set. As this happens, such set treatment becomes more of a textural and less a significant graphic element in the shot, subordinate to the insert. The best production designers will allow for both of these situations in their planning and come up with a design solution that works well for both. One such solution is to use the program logo as a repeated pattern at low contrast across the entire surface of the background and paint it at a higher level of contrast in only one spot, where it appears over the presenter's shoulder in shot. In this situation, the shot remains essentially the same (perhaps zooming out only slightly for the insert), with the electronically inserted graphic box covering the spot in which the logo appears in strong contrast to the background color. Another solution is the use of the low-contrast pictorial background in which the shapes are recognizable in the wider shots, but become undefined textural forms in the closer shots.

Multiple-Head Shots

Life becomes much more complicated when two or more people appear in frame at once. In a dramatic piece, this is somewhat less problematic than for other formats: A sequence of fast cuts between tight head shots or inclusion of over-shoulder two-shots will do the job nicely and have become the staple shots of the daytime soaps. In these cases, the work of the designer is directed much more toward providing believable realistic backgrounds to cover many compositional possibilities and to provide visual depth and dimension for the wider shots. To an extent, this is also true for the talk-show format or for the interview portion of a news or current events program, and often the solution is something like a realistic office, study, or living room environment.

The weakest multiple-head shot is the one in which two or three (or more) news presenters are lined up side by side behind a desk, either handing off to one another or filling the extra time at the end of the newscast

Figure 5.4 Over-shoulder logo and graphic insert

Figure 5.5 Wide shot of the news set, WNBC-TV New York. Design: Frank Schneider

with idle chat. An inherently awkward situation, it is made more so by the ungainliness of the shot: Trying to fit a long horizontal picture into a frame that is only slightly more rectangular than square leaves a lot of dead space top and bottom. Some designers have solved this problem by making the desk an L or a V shape and forcing the group shot to be made obliquely from one side or the other, rather than from straight ahead. Others have filled the dead space with light sources overhead (called *dressing lights* because they're not really needed to illuminate the set) and projected patterns on the floor. Still others pull back very far and show a wide shot including the lighting grid, several cameras, and the floor manager. (Depending on your inclination, you may call this shot either "high tech" or "dirty.")

There are many rules about formats such as the news program (in which the parameters of composition are very tightly drawn) that have been espoused over the years and come to be accepted in some workplaces as law. While these may be quite solidly based in fact (typically, they're the result of trial-and-error mistakes), such rules of design restrict creativity more often than they guarantee success. I remember being told that one should never paint a set green because the human eye could not resolve the flesh tones of Caucasian skin against a green background. This is, of course, patent nonsense, but I've encountered directors who refused to use a shot I'd carefully constructed because they had been told that a talking head should only be shot against a plain background. Whenever you confront such an arbitrary rule, the best course is to ask, "why?" The best rule to follow is not a rule at all, but a question: Ask yourself, "What are we trying to do?" and then analyze your solutions based on the degree to which they work toward achieving that objective. It is never adequate simply to make a design a pleasant composition; you must understand the objectives of the program (indeed, the objective of each shot) as well as the director does in order to tailor your compositional choices in that direction.

Figure 5.6 Background detail, KYW-TV News. Design: Adrianne Kerchner

In general, composition is less critical to the practice of design for television than it is to that for painting or theatre design. In painting or theatre design, the designer has more control over what is seen at any given moment. In television, most shot composition is the responsibility of the director and the camera operators; the best the designer can do is to sit at the director's elbow and advise, cajole, or intimidate as appropriate. The example of the news show is used here because it is the one in which the shots are the most tightly controlled (necessarily because of the need to assemble various other elements into the shot) and are the subject of much more preproduction planning, which must include the production designer. In a scripted dramatic program or a serial comedy, the job of the designer is to anticipate as many situations and camera angles as possible and to achieve (to whatever degree is appropriate) the look of reality on camera. This is especially true of high-end film-style drama series; the practice in these productions is to build all four walls of a room, making two or three of them *wild walls*, which can be disconnected from the others and removed at any stage in the shooting day to make room for a camera. In the theatre-style box set typically used for a sitcom, the tough part is preproduction anticipation of camera positions, so that once the positions are determined, there is no need to stray from the catalog of shots established at the inception of the program.

Whatever the situation, however, the most difficult aspect of composition for the camera is the achievement of a background that supplies exactly what the program demands with exactly the right amount of detail, texture, and control of focus. Making a complicated design is the easiest thing in the world; making a design that is simple and effective is many times more difficult and the skill that sets the best production designers apart from the crowd.

Focus

Directing Focus

If it seems that I refer to *focus* (or, where the attention of the audience is fixed at a given moment) repeatedly and often, it's because it is very important.

Directing the viewer's attention is the most important visual tool available for exercising control. In a medium acknowledged to be principally a visual one, focus is a most powerful tool. Just as the theatre audience is willing to assist us in overcoming its disbelief—to temporarily agree to ignore the other audience members and the physical presence of the theatre building itself—the television audience is quite happy to accept visual cues given to it. Many such cues are governed by the camera, and it's a poor director who does not understand and use the language of focus at all times.

In the simplest terms, anyone facing directly into the camera (that is, into the viewer's face) will tend to take focus from anyone facing away from the camera. Thus, anyone who is being shot over another person's shoulder will, at that moment, appear to be the more important one. This is not to imply that whoever is taking focus will take the stronger position in the viewer's mind, but will have more attention paid to what he or she is saying. If, in an over-shoulder shot, the person speaking is back-to-camera, the viewer will know that the important thing for which to watch is the reaction of the person facing the camera. A twitch of the eye at this point can signal the fact that, unknown to the person whose back is to the camera, something is about to happen (such as a third person entering the scene from behind).

Many other devices can be used to control focus. In a group shot, anyone placed higher than the rest or placed upstage center will tend to draw the interest of the audience. Again, camera angle comes into play: If one actor is placed higher than another, with the camera shooting up at the higher one and the opposite camera shooting down at the other actor, the higher one will tend to appear stronger and that actor's presence will tend to dominate the scene. That is not to say that the one higher up will dominate even if the text indicates a different relationship between the characters, but that the sharp director will use this tendency to support rather than undermine the intent of the scene.

Much has been made of the fact that we read across the page from left to right, and that we are, therefore, predisposed to scan across a TV screen in the same manner. The conclusion, then, is that anything placed to the right of the screen must dominate the picture. There may be scientific evidence to support this conclusion, but it has always seemed to me to be preposterously simplistic and inconsistent with the complexity of human nature. A much more plausible theory is that the eye tends to search for a place to rest whenever it is presented with a new picture, and we can calculate cues to give the viewer's subconscious that will take advantage of this tendency and direct attention where we choose.

Placing a person facing camera and over-shoulder of another gives such a cue. Placing one at the center of a group, all of whom are looking toward that person, will have the same effect. Making anyone in a group look different from the rest (say, dressing someone in a red T-shirt when everyone else is dressed formally) would work well, as would having one person moving erratically in an otherwise static scene (or standing still in the midst of chaos).

Perspective (the apparent convergence of lines over distance) can also be used. Placing someone at the top of a staircase with the converging lines leading up to her, or showing her standing some distance away down a city street will lead the eye in her direction and lead the brain to conclude that its attention is meant to come to rest on her.

Past experience can be used, too. A shot of the same city street completely empty will tap into the viewer's expectation, based on experience, that something will come up that street and will tend to focus attention on the distant end of the street in anticipation of that. Few people can be shown a shot of a railroad track without anticipating the appearance of the train. In addition, people have become accustomed to seeing shots of approaching trains or street traffic from a low angle; therefore, a low-angle perspective shot down a street will encourage such expectation more strongly than a high-angle one.

People unconsciously try to relate the world to themselves and so will look at an empty scene with the expectation that someone is going to enter or will look at a group of people expecting some clue as to which person is the most interesting one to watch. Given a shot of a room, the audience may look to the door for someone to enter; given a gunfight scene in a Western saloon, they'll look to see who in the crowd has a gun drawn and ready to fire. This psychology can be used to great effect: In the film *Tremors*, the last place anyone expected to see an entrance was from the ground—thus the giant worm's unusual entrance had the desired shock value even after its pattern of appearance had long been established.

People have a need to relate to other people, a trait that does not differentiate us from other animals. Cats exist on a plane with other cats, dogs with other dogs. It's pretty basic. Given the classic newscaster shot with the graphic insert, the average viewer will look at the face of the newscaster, scan across to the graphic (which usually helps by identifying the category or context in which to interpret the story) and come back to rest on the face again. We expect to see the face of someone who is talking to us; in the days of radio (when the typical radio receiver was a larger piece of furniture than the modern television set), families would sit around the receiver and look at it. Facial expression tells us a great deal

Figure 6.1 Using composition to control focus

about sincerity and conveys many subtle shades of meaning that vocal inflections alone do not make clear. It should not appear as mystery, then, that most people presented with the image of a person's face speaking to them will tend to look closely at that face to discern those shadings and subtleties of meaning. Even though that graphic insert may be placed on the right side of the frame (theoretically the strongest part of the image), the eye will always tend to rest on the face at the left because people want to relate to it.

This is a big help to the designer of the news set. Remember the compositional problem of program logo size, placement, and contrast? Because of the tendency to want to watch the face in shot, the risk of the background set treatment actually stealing focus from the talking head is greatly reduced. A set treatment needs to jump off the background flat and scream at the viewer to distract from the person in front. Nonetheless, if the background is too strong visually, it may set up a conflict in the viewer's brain that will be difficult to resolve and will tend to turn the viewer off or (worst case) give the viewer a headache. Unless confusion is the desired effect, It is important not to give the audience conflicting signals. (Remember that all design choices have to be made and evaluated in the context of desired effect.)

It is possible to give equal weight of focus to several people in shot at once. The two-handed shot used as a link to commercials in the morning news-and-chat format (and usually accompanied by the words, "...and coming up in the next half hour...") often shows the two hosts sitting side-by-side on a sofa and framed symmetrically, each facing directly into the camera. Attention in these cases usually goes to whomever is speaking and they often alternate bits of text rapidly, handing back and forth at each

Figure 6.2 An equally weighted two-shot

change of subject. The more seasoned talent will have mastered the knack of gazing interestedly at the speaker, even though he or she has seen the script and knows what the other is about to say.

As with composition, the most powerful tool for focus is the camera, with the camera operator and the director having the greatest control. This is not necessarily a bad thing; it demands that the director have some visual competence. (In fact, a rather highly refined sense of composition and focus is usual in directors who have survived any length of time in the business.) A production designer's task can be made much easier when dealing with a director who understands the terminology of design and composition and who can discuss the often esoteric concepts of design without needing a remedial lecture first. The best directors (that is, those who are sufficiently secure in their careers to accept advice without feeling threatened) will typically consult with their production designers on matters of shot composition, and some who can afford the fees will ask the design staff to produce a complete book of storyboards that include each anticipated shot. A close alliance between the designer and director results in the most effective finished product, since little is left to chance or until the last minute.

Single Camera or Multiple Cameras

At the high end of television production, in which the medium is film (or, in Europe, high-band PAL videotape), the project is often shot with a single camera, *film style*. The need for strict shot continuity, to make the shots match when they're edited together (perhaps weeks later), is essential. Continuity errors, such as an actor who has his hand in his pocket in the wide shot and instantly has it on the other actor's shoulder a fraction of a second later in the close-up, can destroy scene credibility and steal focus from the intended focal point with disastrous effect. Situations such as this demand that shot composition and focus be hammered out in fine detail to avoid wasted time and very expensive reshoots. When a complicated scene must be shot again, at a cost of thousands of dollars, the value of a careful and thorough production design staff and a close working relationship between the design department and the director becomes tangibly evident.

In the typical television studio, operating with multiple cameras and much smaller budgets, storyboarding of every possible shot may not be feasible. One solution is to agree on a limited vocabulary of shots and guarantee that the show will be restricted to those and only those. A good example is the main set for "Cheers," which has adhered strictly to the box-set formula, never shooting over-shoulder from the bar and showing us the fourth wall. In fact, the cameras never cross the imaginary proscenium line across the front of the set: The left camera will truck around the bar far enough to shoot down its length to Norm's spot at the far right end, but never goes far enough into the set to actually see the upstage side of the bar. There are certain preestablished focal points on the set at which principal characters are placed, and to which most of the critical tight groupings of actors are typically restricted. Places such as the entrance door, either end of the bar, the two tables at the near corners of the bar, and the

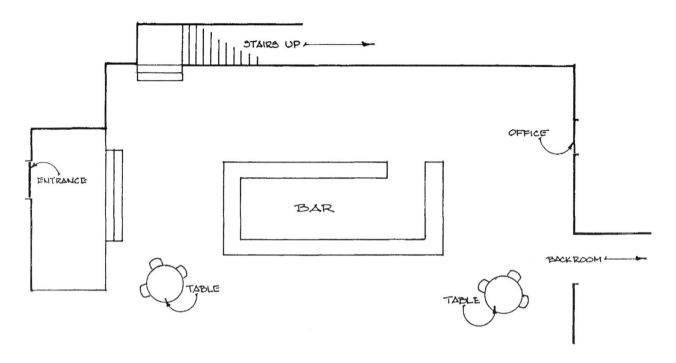

Figure 6.3 Schematic of "Cheers" set

door to the manager's office have been established as focal points that are easy to shoot with the typical three-camera studio setup and around which actors can be placed to facilitate group shots and over-shoulder shots. The planning that went into the layout of the set was extremely thorough. The set has lasted many years without need for any revision and has accommodated a great variety of shots and compositional situations. These places are all psychologically natural points of focus in such a room. An entrance to a room is always a point of focus and the bar in a saloon, since it is a point of congregation and activity, is also a natural point of interest, based on the innate interest people have in the doings of other people. With both the original cast and the second cast, the office door was a place at which the sexual tension between Sam and Diane and between him and Rebecca has been played out. All of these focal points have been placed in such a way that a group of other people can eavesdrop, comment upon, or easily join in with the action being featured at any given moment, making the director's task much easier by facilitating the sort of placement of actors and construction of scene dynamics for which the script so often calls. While it may not provide the sort of visual variety typical of a feature film

---

Unlike paintings or still photographs, focus for television is more than just a function of composition. The audience will focus on the places from which they expect important actions to come. These may be obvious expectations, such as the far end of the road or the door to a bar. Or they may have expectations learned over successive viewings of a particular program, as is the case with Norm's corner of the bar in "Cheers."

with a much larger budget and only one camera to accommodate at any one moment, the set for "Cheers" is very functional in terms of direction and action and reflects a great deal of preproduction attention to the script requirements and a production designer with a good understanding of the possibilities and limitations of the three-camera studio setting.

Focus may be based on psychological factors or on somewhat more direct physical ones. As with the theater, we can control focus by not permitting the eye to see anything except that which we want the mind to focus on. We can do this simply by zooming into the scene so tightly that nothing else fits in the frame, a technique used often for intense emotional scenes in daytime soaps. We can use a shallow field of lens focus (using the word in its other meaning) by having, for example, a person in the foreground sharply in focus while another is out-of-focus in the background. By changing the focus of the lens, and bringing the person in the background sharply into focus (while losing focus on the foreground person) we can control the focus of attention absolutely. Additionally, we can use the cruder (but no less effective) theatrical tool of lighting only that which we want seen and letting the rest of the frame fade into black.

Returning to the more restricted composition of the news set, the careful lighting of a talking head can do wonders for control of focus. Well-placed lighting equipment and light sources, the intensity of which has been carefully balanced to achieve the maximum three-dimensional appearance, will bring the face away from the background and make it more interesting to look at. This technique, used extensively in soaps, uses a strong and usually fairly intensely colored backlight on the top of the head and the tops of the shoulders to frame the face and pull the whole torso of the actor forward and away from the rest of the information in shot.

Conversely, withholding information can peak the viewer's interest and thus direct attention to the part of the shot from which the critical information is missing. The classic *incognito shot*, often used in documentary programs, with a talking head shown only in silhouette, will command attention more strongly than a well-lit face in the same shot. People need and expect to see the face of a person talking to them, and denial of this visual reinforcement is enough of a tease to reliably ensure concentrated attention for rather extended periods.

There are as many ways of controlling focus as there are creative people planning and shooting video. The format most reliant on creative control of viewer attention is the commercial; given the amount of impact that the typical commercial spot has to make in order to sell its product in 30 or 60 seconds, the ability to do so has become a highly refined skill and its practitioners a rather select group of artisans. A study of the best commercials (usually, but not always, those produced for national network distribution) is an excellent way of understanding the visual power inherent in television and gaining an appreciation for the creative skills applied in mastering the compositional and psychological tools that focus the attention of the viewer where the director wants it.

7 Dynamics

The Use of Motion

Of all of the areas of television production, the use of motion is the area in which the designer has less input than do other members of the creative staff. *Blocking*, or movement of the performers through the set, is one of the important functions of the director, and camera movement is usually decided between the director and camera operator. Many production designers opt out of this phase of creative planning and go off to dress the set for the next scene. (Unfortunately, time constraints often make it too inefficient for the production designer to hang around the set throughout the shoot.) Artistically, it is best for the designer to participate at every stage, but television being a profit-making business in the United states. makes that the exception rather than the rule. In many of the European production facilities, it is standard practice to have the production designer present on the set during all shooting, assisting the director in setting up shots and with all aspects of actor blocking and camera movement. It is fair to say that this is typical only in the state-supported TV studios in which cost-effectiveness is not the overriding consideration.

Nonetheless, the designer will usually be involved in all of the preproduction planning, whether in commercial or noncommercial production and so will be present when the more important decisions about dynamic structure are made.

In the most basic way, the designer must know well in advance what sort of movement the cameras will make in and around the set. If a camera is to shoot through a doorway, where will it be? How wide will the shot be? Will we need to put a rug on the floor? Pictures on the walls? How far around the door opening will it see? All such considerations must be addressed in advance. It is expensive and wasteful to have a studio full of actors and crew sit on their hands while the carpenters attach an extra flat and the painters match its color to the others and wait for it to dry.

Designing Backgrounds: Putting Motion in Context

If a director describes blocking in which an actor opens a door and walks into a room, the designer will need to ask where the camera picks up this entrance; whether it follows the actor in, tracks around her, shoots her wide or tightly; and where the action (both actor and camera) finishes. All of this is critical information; not only must the background of the shot be covered at all times (that is, the camera must not see off the set) but the texture of the background must be taken into consideration to decide what sort of image it will project as the camera moves across it and how that will affect both the static shot as the actor enters and the shot at which the camera comes to rest when she stops moving.

The long tracking shot is the one form of camera movement that has the greatest impact on the production design. The set must be planned to accommodate the camera dolly (or the camera track if the scene is being shot outside the studio) without permitting the passage of the dolly through the set to be apparent in shot. In addition, the layout of the set and the placement of things in shot must be paced in such a way as to appear at the correct moment, based on the speed at which the camera and the actor moving across the set are traveling. Jim Cartwright's "Road," produced for the British Broadcasting Corporation (BBC), contains a superb example of such planning: A man walks at a brisk pace along a street, passing block after block of deserted brick row houses. The pace at which he walks, the camera matching his gait, and the rhythm set up by the passing of each side street perfectly counterpoint the rhythm of the text as he talks aloud to himself. This kind of visual poetry is sufficiently rare to lodge itself firmly in your mind whenever you see it, especially when you become aware of the degree of subtlety and creative cooperation required to achieve it.

Moving Cameras

It is difficult to separate dynamics from composition and focus, because the psychological processes are closely intertwined. Much of the process of focus is the determination of the point at which the eye comes to rest; the motion of the camera (and the direction of that motion) will determine that place where the eye comes to rest. As such, motion is a powerful tool in control of focus. Just as the shot of a railroad track anticipates the coming of a train, a shot from a camera placed on the front of a speeding train anticipates those things that are out of sight beyond the next curve or at the far end of a tunnel. Thus the choice made by the director, camera crew, and designer to mount a camera on the side of a police car to shoot a chase scene, for example,) is a compositional choice in that it fills the frame with the road and the front of the car. It is also a focus choice in that it leads the audience to focus its attention very closely on a point down the road in anticipation of an oncoming vehicle or the vehicle being pursued and it is a dynamic choice in which the road surface and the buildings alongside it become textures and swatches of color in the passing blur.

In any tracking shot, focus is automatically given to the figure the camera is following and is taken away from the things in the background that are swallowed up in the motion. If we're watching someone driving down a highway from a camera fixed to his car, we will tend to focus on the person in the car rather than the passing traffic. However, if one of those cars pulls alongside and matches its speed to that of the first car, our attention will shift to the new object that is now moving in synchronization with the camera and therefore, part of our scene. Typically, we might expect some important action from that second car, such as shouted information, a facial reaction indicating a relationship between the two drivers, or perhaps a gunshot.

Of course, it is hardly necessary to clamp a camera onto a car or train to achieve a dynamic state. Technically, any motion may be classed under this heading and may certainly include something as simple as a *truck right* or a *zoom in*. Everything the camera sees is the purview of the designer, and the designer often has provided input to the director regarding the simplest camera dynamics. At the most basic level, the means of getting from one shot to another has dynamic considerations, and a designer may want to know whether the director plans to cut, pan, or truck from one shot to another to be able to plan the texture and finish of the set accordingly. If a camera shooting a talk program is required to pan across the set (following a mobile host, for example), the designer may want to avoid a set treatment with many vertical lines, which may create an annoying strobe effect as the camera pans across. You may note that the portion of the typical news report in which the weather person stands in front of the map is always a *cut* from the previous shot, even if the weather reporter was in that shot. Why not simply follow her across? Because the weather map is nearly always electronically inserted (where the audience at home sees the map, there is really only a solid blue or green panel) and the illusion would be spoiled if the camera panned across it and revealed that the map image is positioned in sync with the camera shot rather than a part of the studio news set.

Dynamic Texture

A very powerful dynamic set element—so strong an image that it has become the trademark of the program and the central figure in its opening graphics—is the wheel used on "Wheel of Fortune." The dynamic attraction of the wheel's motion and the slower (but psychologically very powerful) action across the puzzle board as the words are revealed are so strong that they have become the dominant shots, only occasionally giving way to shots of the host or the contestants. The contestants are in shot much less frequently than on most other game shows—and usually only when the wheel and the puzzle are both static: At the beginning and end of the program, and just prior to commercial breaks.

We would be very much mistaken, however, to confine the consideration of dynamics to those scenes in which the camera is moving. A program has a dynamic pattern when seen as a whole, and the counterpoint provided by moments of stasis are as significant a contributor to the overall dynamic texture as the visual excitement provided by those moments in

---

Truck is to roll the camera left or right.

Zoom is to use the internal optics of a zoom lens to make the subject seem larger without actually moving the camera closer.

Cut is an abrupt change of shot, unlike a fade.

which the camera is moving. An excellent example of such a carefully constructed dynamic texture is the Wolfgang Petersen production of *Das Boot* (*The Boat*) in which moments of tension and panic-level excitement are carefully contrasted with scenes of crushing inactivity. Petersen made extensive use of the Steadicam, a device that permits a camera operator to walk or run while holding a camera, and yet produces a shot without the debilitating amount of camera shake inherent in any hand-held shots attempted without the aid of such a device. With it, he was able to get some superb footage of a sailor's-eye-view of a run from one end of a submarine to another, ducking through hatchways and dodging obstacles along the way.

Television is a two-dimensional medium. As such, its design considerations are very similar to those of painting or graphic art. The principal difference (and it is a very big one) is that TV can move. The motion of the camera and the motion of things in shot are tools of the medium as much as color, line, and form are to the art of the painter. As a production designer, your craft requires that you master the subtleties of the medium's dynamics as well as those of composition, color, or focus. Make this a part of your palette and the process of creation will become one of cooperative gestation with the other creative personnel and the payoff in personal satisfaction will be greater.

8

Color

The History of Color

Perhaps the one development in the history of the medium that had the greatest impact was the introduction of color. Film made its big splash of color in the early 1930s with *The Wizard of Oz*; it was another 30 years before color became standard in broadcast TV programs and another ten after that before the networks stopped hyping it. Walt Disney's prime-time weekly program was titled "The Wonderful World of Color" and NBC adopted the peacock with the colorful array of tailfeathers (which it still uses) to tout its color broadcast capability.

The reason it took television so long to incorporate color into its product was that it was inherently a more difficult task. Having devised a film stock capable of producing a color print, it is not necessary to junk the entire stock of equipment to make and show movies. Because of the necessity to encode and decode picture information to create television, it was necessary to throw away great quantities of equipment at both ends of the system. Studio cameras, switchers, cables, and home receivers all had to be replaced for the transition to be complete. To avoid chaos, the new color system had to be fully compatible with existing black-and-white TV sets, to permit those without the financial resources to replace their sets to continue to participate in the medium. This resulted in a compromise system; the opportunity to increase the resolution of the broadcast picture and to produce a clean and consistent color palette was missed in order to avoid the technical headaches and economic disaster that would have resulted. The result is that we are left with a picture that not only has poor definition of visual detail but has some downright maddening color artifacts. These are mistakes made by the decoding system, which tends to confuse brightness information with color information. To fully appreciate the deleterious effect of this, take a trip to Europe and watch a program on the PAL system, which is (relative to NTSC) virtually artifact-free.

NTSC's idiosyncracies can be frustrating for the designer. Jules Fisher, the famous theatre and television lighting designer, used to complain about lighting for television, noting that with the NTSC system he could spend weeks planning the lighting design for a scene only to have a television viewer turn the color knob on the TV set, thus negating all of Fisher's work in a single stroke. It is true that this also mitigates against the successful use of a subtle color palette, because final control does not rest in the hands of the program's creators.

This need to make all succeeding systems compatible with the standard that was adopted in the late 1940s will likely also be the demise of HDTV. The industry will not soon discard all of its broadcast equipment and the government will not soon reallocate the broadcast bands to accommodate a true HDTV system. The talk now is of a system called (ATV), Advanced Television, which would provide somewhat enhanced picture quality while remaining compatible with existing hardware. Whether the various artifact problems and other color problems which plague the present system will be solved remains to be seen.

Designing in Color

There is a system in place for the control of color balance, and among video professionals the maintenance of consistent color quality is usual practice. It is standard practice to set up a video system before shooting anything by generating electronically produced color bars and using the vectorscope and waveform monitor to verify the accuracy of the line monitor from which aesthetic color decisions are made. Setup of the system color balance is typically done by engineering staff, and is not the responsibility of the designer; however, it is a good idea for the designer to understand the process and be able to verify firsthand that the system is functioning as it should. This is equally important to the lighting designer and, in some situations, such as that in the English system, the lighting staff actually has control over the color and luminance adjustments of the cameras. In most cases, an experienced eye will easily detect a color mismatch between sources without the aid of a vectorscope. One of the best indicators of color mismatch is in flesh tones or in areas of white in the shot; slight tints will often show up in these cases, and asking the technical director to cut rapidly between shots will allow direct comparison and will show up any color differences.

Having adjusted the cameras to give both satisfactory and consistent color response, we now record those color bars at the beginning of the tape. This will establish a reliable reference for those whose job it is to play the tape back to ensure that the color palette the production designer chose will be reproduced accurately. Color bars also include bands of black and white, used to evaluate and adjust the white level (peak output) and black level (pedestal) of the system. Many cameras now have intelligent circuitry built in, which automatically performs the white and black balance functions at the flick of a switch.

Over the years, video cameras have become more sensitive to low-light situations and their contrast ranges have been expanded. While the contrast range has been expanded from 20:1 to 30:1, it is still much less sensitive than film (at 100:1) or the human eye, which can detect subtle differences of brightness not reproducible in any of the available media. Similarly, video cameras have become much more color responsive and better able to reproduce slight color variations. Nonetheless, there is still a marked tendency for the system to flip-flop and to try to interpret areas of slight variation in contrast or color difference as much greater than they really appear.

The property of color that defines the color itself (the thing that differentiates red from green) is its *hue*. The intensity of that color (say, the difference between red and pink) is its *saturation*. In general, a video picture tends to show colors in greater saturation than they would appear to the eye. Try shining a light on a white surface and framing it up in shot, then hold several relatively pale colors of lighting gel in front and observe the effect on camera. You will notice that the video reproduction process involves a great amount of interpretation of color and that, in addition to increasing the saturation, it also involves various shifts in hue. It would be convenient, I suppose, if some engineers were to conduct a series of laboratory tests and produce a chart or a mathematical formula by which to compute the color alterations produced by video cameras. These cameras, however, do not all react consistently and cannot, therefore, be quantified uniformly. As they become increasingly reliant on chips rather than tubes and thereby require less adjustment, their performance also becomes more consistently predictable to the production and lighting designers, and life becomes easier. There is, as they say, no substitute for experience, and you will eventually develop an intuitive ability to adjust colors to compensate for the distortion inherent in the camera. Until that time, it is wisest to religiously do color tests on camera before committing to any palette. Test your color choices by painting pieces of plywood at least a foot square and getting samples of wallpaper and fabrics of a similar size and arranging them on camera and under studio light before committing to them.

White belongs in the color spectrum, since it contains all colors, and we'll include it here. Most television systems cannot accept pure white. Typically, they are adjusted to accept reflectance no greater than 80% of peak white and any greater signal will burn. When this happens, the object in question appears to glow very brightly, rather like the flame of an arc welder. This is not only an extremely annoying effect; it will often permanently damage tube cameras. The effect known as a *comet tail* results when a tube camera is pointed at a light source or a highly reflective surface. This is characterized by a red streak as the camera moves across the scene, hence the name. Tube cameras have a memory and ,when left pointed at such a source for a long time, will acquire a permanent red spot that cannot be corrected. Chip cameras do not have such a memory, but still produce the unpleasant effect of flaring up when pointed at any too-bright surface or directly into a light source. It is incumbent on the production designer to be aware of this, and to avoid any peak-reflectance surfaces on the set or clothing of the talent, and to avoid a set design that requires that a camera be directed toward point sources of light.

Many paint supply houses that cater to the TV industry routinely supply shades of grey in degrees of reflectance. Probably the most popular of these is the 60% grey, which is virtually guaranteed not to flare, yet is sufficiently reflective that it can be washed with colored light and give the effect of a surface painted in that color. This is a good, reliable all-purpose solution to the question, "What color do I paint it?" Many a variety show has used this system to good effect, especially combined with one or more stripes of silver tape stuck on around the edges of the platforms. In my opinion, this is a cop-out, because it leaves the lion's share of the color choice to the lighting department, and completely falls apart when a white light source (such as a follow-spot on the talent) is present, as it will wash out the color.

Color is a powerful tool in this medium, and the advent of color TV with a modicum of subtlety has greatly expanded the scope of the production designer. Few visual cues have more direct psychological impact on the viewer than color; in today's visual vocabulary, a black-and-white film will typically lead its audience to expect a serious treatment or cue the audience that the story is set in the past. In the days before color TV, the intensity of the lighting was the major visual cue as to the seriousness or frivolity of the piece; a comic piece would be lit flat and bright, a serious one darker and more sculpturally. These conventions still persist: Soaps, and some prime-time dramas, are often lit with strong shadows and strong backlights, while comedies are lit brightly and indiscriminately. In the subtler vocabulary of color television, the overall saturation of the scene and the relative contrast of colors within the palette are cues to the viewer. Lighting for a game show will tend to be more saturated, and the colors more often primaries or secondaries rather than subtler tones. A show such as "Jeopardy," with its bright blues, oranges, and yellows told us within milliseconds of first glance that we could sit back and rest our psyches. One of the more serious dramatic programs, such as "Law and Order," will tend to use a more restrained and naturalistic palette (naturalistic, that is, to those who don't live in fast-food restaurants). This is one reason why many of the dramatic series are originated on film: The film medium is inherently less inclined to oversaturate color than is video and is capable of conveying subtler shadings more accurately, even if it is being broadcast over the television system. The HDTV system gives a picture more like that of film than conventional TV, though still not quite as good; the trend in technology, however, is in that direction, and it is not unreasonable to assume that a day will come when video gives us a color reproduction as subtle as film.

Color, Lighting, and Camera Technology

Again, the integration of set and lighting designs can't be overstressed. Because the colors you choose to put in the shot are at the mercy of both the lighting and the cameras, it is a wise designer who will take the time to comprehend the intricacies of these media and to plan accordingly far enough in advance of the shoot to discuss her desired effects at length with the lighting and engineering staff and to involve them in the creative process. (One of the most valuable skills a designer can acquire, and one that is rarely taught in school, is the ability to interrelate with other creative and technical people and to involve and lead them through the production process.) Often, these colleagues may have more experience than you and will come up with technical solutions to aesthetic problems, which will either bail you out of a difficult situation or spark an idea for a look that you may not have thought of or may have thought impossible. Adopting the correct attitude—one of informed co-operation—will get you much further than adopting a hard-nosed and dictatorial stance. Television, like theatre, is a cooperative art, as you will see.

The most important object in shot is nearly always a face and, in general, this is your (and your audience's) most usual point of color reference. Most of the time, television shows us faces, whether the program is the evening news, a talk show, or a drama. Stories interest us because they're about other people and the words and actions of people within the context of the story are what advances the plot. Thus it is that much of your time will be spent seeing to it that the face in shot at any given instant looks good and is framed and set off well.

In the case of the news program, it's important to pick a background tone that recedes from the face of the newscaster rather than blending with it. Thus, flesh tones and tones that come too close to flesh tones on the grey scale tend to flatten the face into the background and work against the intent of the shot. The darker the complexion of the head in shot, the further down the grey scale goes the color range that we must classify as flesh tone. Because most stations now employ newscasters from a variety of ethnic backgrounds, the range of these tones can be very broad. Here is where your lighting designer can help; typically, a dark-complected person will be lit more intensely and may have a stronger backlight, and one with a more saturated color in it, to frame the face and hair and provide more reflective highlights from the cheeks, forehead, and chin. All of these tricks help to define the face and shape the hair to bring the head forward and make it seem more three-dimensional.

This sort of help from lighting will make a huge difference if you have a light-skinned, a dark-skinned, and an in-between-skinned assemblage doing a program in front of the same set. You could shoot the light complexion in front of a black drape and get a nice effect and you could probably do so with the middle-toned face but the very dark skin will simply vanish into the void if you put that person in front of a very dark background. This is when you curse the video contrast ratio. Unfortunately, cursing rarely improves the shot and you must find another background color or a pattern involving a variety of background shades that will set off all three faces nicely. This was a much more frustrating problem in the days of monochrome TV when color differences could not be enlisted to help and the contrast ratio was only 20:1. Today, it's much easier, because in addition to all the shades of grey, you have the whole spectrum of color to assist you.

The Psychology of Color

Finding a good contrasting color to set off the face is often not adequate. The brain perceives colors in a relative way and will tend to shift its color perception depending upon the predominant color in the frame. Thus a light-complected face in front of a bright red background may tend to look a bit green (not a desirable effect) and a very deep-brown face shot against such a saturated red may tend to look blue-black or deep blue-green (also not a healthy look). I know of a station that uses a graphic background, inserted electronically, which is a very rich greenish-blue. This has the effect of making the faces in front of it shift toward the complementary colors, red and yellow, and take on a distinct look of jaundice. The choice of this color is wrong for two reasons: (1) the color shift it causes in the faces by its sheer oversaturation and (2) for the frivolous effect it presents subliminally to the viewers, placing it in the same palette as, say, "Wheel of Fortune" and thus inclining the viewer to interpret the news in the same context as he interprets Vanna White.

A common and very practical solution to the news-set color problem is to choose a neutral color, such as a desaturated blue, for the general background. This may be achieved by moving the shade up or down the grey scale—by the addition of white or black to reduce the content of pure color. Having done this, portions of the set can then be highlighted—patterns, for instance, or a program logo—to add interest and break up the background area, as long as they remain a minority portion of the total area in the shot. It is the predominant color that the brain perceives that sets the reference within which we evaluate the other colors we see; therefore, by keeping the background color less saturated overall than the color of the face in front of it, the face will be what dominates and our perception of the background will adjust to the color of the presenter's skin, rather than the other way around.

The dramatic program is inherently an easier lighting situation. There is typically only one point of focus in a scene; it is not necessary to see a station logo over the shoulder of the presenter and therefore, the background surface (whatever its color) need not be lit as brightly as the face in the foreground. This is another way of reducing the saturation of background color: Withholding illumination from a surface will have the effect of moving it down the grey scale, much like adding black paint would do.

The human brain has difficulty resolving saturated colors in close juxtaposition and the video camera seems to aggravate this difficulty. The effect is most noticeable in the case of red and green. Two solid blocks of these colors adjoining each other will be difficult to resolve, and the line where they meet will seem to the brain to shift and move. The net effect of this, over even a short period of time, will be a headache. Obviously, it is important to avoid visual effects that are liable to cause our audience to switch off, so the production designer must avoid such a situation. I've seen a few shows (usually children's programs or game shows) in which the clash and sheer intensity of the colors were enough to make me change channels while the opening graphics were still on screen.

Taste and restraint would seem to be the watchwords for color choice, but it is yet another area in which such things are infinitely more easily said than done. Some people seem to have an easier time than others in working with color and appear to have no need for guidelines or geometric color models on which to base their choices. For those who need theoretical support and formulas to assist them, many books have been written that deal at length and in depth with the subject, and may indeed help. It has been my experience, however, that color palettes are trendy and certain combinations that are thought to be the ultimate of taste and sophistication this year would have brought people to the brink of nausea five or ten years before. Our taste is conditioned by what we see around us daily and these things change constantly. Pink cars, for instance, are quite uncommon, as are pale green or purple. In the 1950s, though, they were among the most popular colors. Rather than expound on theories of color combinations, then, I would encourage you to get in touch with your surroundings and develop an awareness of the color palettes with which you come in contact everywhere.

If you're doing a kids' show, go to toy stores and look at the palette that is used there: Study the packaging, the toys themselves, the children's books, and you will begin to get a feeling for the combinations that appeal to today's generation of children. Similarly, designing a dramatic show dealing with, say, the 1920s should prompt you to dig up books and paintings from that time that will give you a sense of the color palette that those people regarded as pleasing. Taste in color combining is a relative (and very fluid) thing and, rather than attempt to set down any rules and guidelines, I prefer to point you to the more difficult course, but the one that will serve you better in the long term. As with any important design decision, choice of a color palette should begin with the question, "What is the objective?" Having established an answer to this as a starting point, you may now move into the areas of theory that do apply: the psychological effects of colors, the seriousness (or frivolity) associated with certain colors, and the ways in which they may clash with or accent those colors that you cannot alter (such as skin tones). As you research the background and content of the piece, its characters, and its audience, you will narrow the field until the choice becomes rather easy. Even with open-ended programs such as talk shows, you will probably discover color preferences held by the show's host or by the director, which will lead you in a certain direction. Remember that your function as the production designer is to support the intent of the show; imposing a system of color combinations that pleases you or conforms to some theorist's arbitrary guidelines may very well fail to reinforce the tone of the program and thus may be the wrong choice. I recall seeing a program about sewing and needlework, which was aired on weekday afternoons, and was quite obviously aimed at the housewife market. The set was painted in pastels (blue, pink, and yellow) and had many accents, such as window curtains, with delicate floral print patterns. I strongly dislike the pastel "country" look, but to the Midwestern housewives to whom the program was directed, it was perfectly appropriate. Similarly, the lineup of Sunday morning news and business programming tends toward subdued colors and the deeper wood tones, to recall the surroundings found in corporate offices and boardrooms, and thus appeal to the people who identify with those surroundings (who are, presumably, the desired audience).

Paints and Pigments

Since I'm addressing you as a production designer and not a scenic artist, I won't go into the theory of pigment mixing in great detail. However, it will do no harm to understand how colors are made and how they interact with each other. Unfortunately, the rigors of budget and time dictate that most television design involves off-the-shelf wallcoverings, paints, and finishes rather than the made-from-scratch stuff that was common to the theatre trades. It also dictates that the designer working in TV rarely has time to do the elegantly detailed painter's elevations once routinely presented to the paint shops.

If you've not done so yet, find as many of the books dealing with scenic art as you can and study them thoroughly. They will explain the color-wheel model of the color spectrum and the various names given to color types at various levels of removal from the three primaries. Theoretically, any color can be created from the primary colors: red, blue, and yellow. However, any seasoned artist will tell you that, while this is theoretically possible, it's well nigh impossible in many instances. Nonetheless, understanding that most colors are made up of varying portions of other colors makes the essential task of color analysis quite easy. Eavesdrop on a design conference and you may hear a shade of red described as "having too much blue in it," "favoring the yellow," or some other such criticism. Learn to think of color in these terms, and learn the theory of complementary opposites. There will be times when your design brief has no peg on which to hang: when you are doing a game show or a variety show and have no reference in the real world from which to draw your inspiration. In these cases, you will often be required merely to find colors that make pleasing combinations (as defined by the current trends). An ability to analyze, mix, and juxtapose colors skillfully and according to the applicable theory will get you out of a tight corner.

My best advice is to urge you to build up a good selection of gouache paints and learn to use them. Find out how a wash of a complementary shade will make a more effective shadow than will a wash of grey. Discover how you can make paint look flat or give it depth with blending, spattering, and washing techniques. Even if you don't need to paint scenes in your design work, do it in your sparetime; the more you do, the better will become your color instincts, and the more subtly you will use color in your design work. It's a sad fact that most people designing for film or TV spend precious little of their time in the creative process.

9
Mechanics

The Film Model

I've made no secret of the fact that the television and video medium has evolved from film and theatre (more so than many of its practitioners like to admit); and I'll go a step further and assert that, mechanically, it is just like film.

The sense of motion in cinema is an illusion. In reality, as most people know, the motion in a cinematic film is produced by the rapid (and sequential) flashing of a large number of still photographs in rapid succession. There is no real motion in film; the apparent motion is produced in the mind of the audience as a result of the brain's inability to register each still frame separately.

A strip of film, then, consists of a sequence of still photographs covering most of the surface area and, alongside, a magnetic stripe on which is the sound track (synchronized with the pictures). As the film is played back through the projector, a gate opens and closes rapidly (24 times per second), flashing each frame briefly and moving on to the next.

Editing film couldn't be simpler; we simply cut the film with a very sharp knife and tape it to another bit of film, assembling the pieces until the film is complete. (Such an edit is known, logically enough, as a cut.) More complicated and subtle edits, such as dissolves (or fades), have to be done optically and are much more costly.

The Videotape Recording System

With video, the essential concept is exactly the same: A videotaped piece is also a string of still frames, each of which reveals in succession a bit more

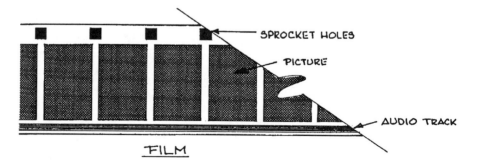

Figure 9.1 Film formats

of the action. Because the system is electronic rather than photographic, the speed with which the frames fly by is synchronized to the number of cycles per second (CPS) of the standard power supply. Thus, in the United States., which has an electric supply that cycles its pulses at 60 per second, the standard rate of frame playback in American TV is 30 cps. It is 25 frames per second in countries that use 50-cycle power supplies. Anyone who has done video editing will be well aware of this, since it is possible to slow down the tape on an editing machine sufficiently to see each frame go by individually.

The big difference, of course, is the fact that you can't hold a piece of videotape up to the light and see the pictures on it. These are recorded magnetically, like the sound track on a cinematic film, and can only be decoded by a video playback system. How is this done? Both systems begin and end with essentially the same things: a real-life three-dimensional scene at one end and a two-dimensional representation of it at the other end. While it is enough for the film system simply to focus the light from a scene on the film stock to create an image, the video system is a bit more complicated.

A video camera uses a lens that is optically the same as a film camera. Where the film camera would focus the image on the film itself (this is referred to as the focal plane), the video camera focuses the image on an electronic receptor. This receptor can be a tube or it can be a charge-coupled device (CCD) or *chip*. The system of electronics inside the camera (called the *head*) scans the tube or chip at 60 cps (50 in Europe). Each scan makes up a field and two scans make a complete picture, or one frame. Hence, there are 30 frames per second in the NTSC system and 25 in the PAL system. In the higher quality systems used for high-end industrial or broadcast applications, there are three receptors, rather than just one, in each camera. This results in better color accuracy and better resolution. At the output end (that is, your home TV), the system simply reverses itself. The television receiver or studio monitor contains one or several electron guns that fire at the screen causing it to glow. A monochrome system has one gun, a color system three. These are aimed by an electromagnetic field and are synchronized so that they scan horizontally across the screen tracing layers of lines that make up the complete picture. This must be done in perfect synchronization with the original camera head that encoded the information in the first place or your picture will not make visual sense. Technology has achieved this; we all know that it works without having to know

Figure 9.2 Videotape format

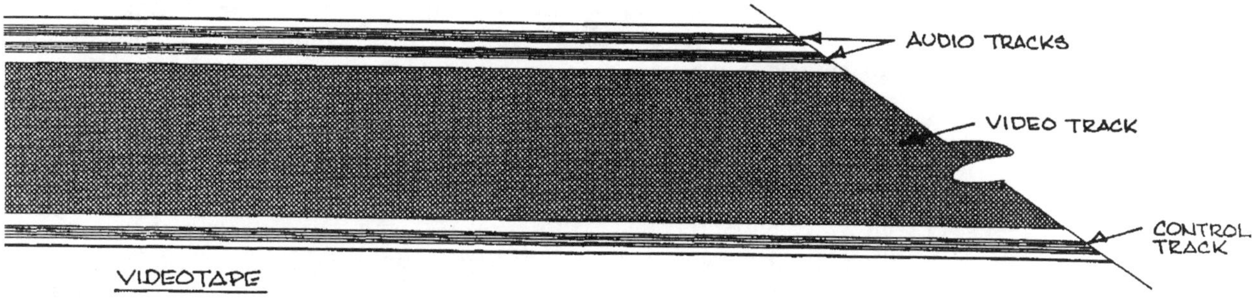

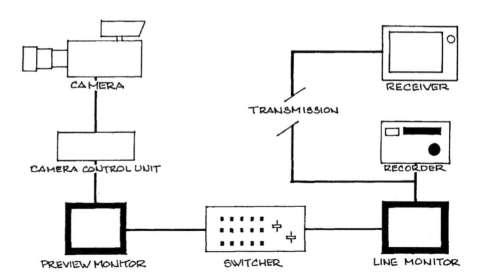

Figure 9.3 Schematic diagram of the video system

precisely how. Parallel to the encoded picture information is a sound track (or two) and other tracks that carry information that is used for the identification of frames and for image synchronization. It is a very complex system, and it carries a great deal of information besides that which is required to make up the pictures; this extra information area is called the *sideband* and has no counterpart in film other than the sprocket holes that cause the film projector gate to open at the right moment.

Advantages of Electronic Imaging

The beauty of the video system is a by-product of the fact that its information is electronically encoded. When you shoot a photograph (or several thousand photos as you do when making a film), the image is fixed at the instant of exposure. In order to alter the image in any way (for example, to fade out to black or fade from one image to another), you have to rephotograph the sequence through special (and costly) equipment. Because the pictures in video are converted to electrical impulses and then reconverted to impulses of light (and a coherent picture), we can do things to those pictures that we cannot with film or with an ease and speed that we cannot with film.

Just as music has been revolutionized with the introduction of sophisticated electronics, so has video grown. The switcher (vision mixer in Europe) has become a complex and sophisticated machine, roughly parallel to a music synthesizer. As its name (in U.S. parlance) suggests, it was originally a switchboard that permitted the technical director to make abrupt cuts from one camera to another. As electronics progressed, it became possible to do instantaneous fades (dissolves) from one source to another simply by pulling a small lever. Later, it was expanded with the wipe function, in which a line traveled across the screen and thus effected a change of picture. Soon there were dozens of wipe options, each taking a different geometric shape. More recently, there have been *pixelization wipes* added, in which the picture breaks up into thousands of little dots and dissolves, or *cube wipes,* in which the pictures appear as different sides of a cube or other three-dimensional form. The more sophisticated hardware now allows user programming of these functions: If you can invent it, you can do it on-screen.

Special Effects

Because electronic signals can be mixed, faded, and superimposed instantaneously, television has been able to create in real time complicated visual effects that film could not hope to duplicate with the same speed or at such a low cost. It has become routine to superimpose graphics over live pictures, put pictures from several sources (often thousands of miles apart) on-screen at the same moment, and key complex graphic constructions into a shot with a live presenter.

A chroma-key unit, or color-separation overlay (CSO), substitutes a signal from a second source, such as a weather-map graphic, wherever it sees a certain color (typically a specific shade of blue or green). Therefore, when the weather person stands in front of a blue flat, the system inserts a weather graphic in place of the blue. The person appears to be standing in front of a giant map. As long as the person is not wearing the same shade of blue, the effect will work. However, if he is, that part of him will appear not to exist.

The trend in switchers is toward more options and more sophistication in their design so that the high-end products now offer the ability to make many images appear on the screen in an infinite number of arrangements, and to make them come and go by the most amazing acrobatics. The function of the machine, however, has not changed: It is there to get you from one source to another.

The Impact of Computer Technology

Many of the complex effects now built into the switcher were formerly the purview of the postproduction suites. All of the fancy effects, and even the most rudimentary electronic character generators (video type-font machines) are a product of computer technology and have progressed in tandem with developments in the computer markets. As computers have attained gargantuan memory capacities, television-effects systems have become faster and more flexible. Not many years ago, video effects (while much cheaper and faster than optical film effects) were much less than instantaneous, taking as long as a half hour or an hour to complete. There are character generators still in use which drive their operators crazy with the phrase "rebuilding, please wait." These were top-of-the-line machines some years ago; now, they are less efficient and slower at these tasks than many home computers. Thus, the trend in video effects has been a move from the postproduction suite to the on-air control room.

There are still many postproduction suites, and they're busy. In the 1970s, an English company called Quantel marketed a new gadget called the Paintbox. It boasted a mammoth memory capacity and a rather large price tag, but many of the larger television facilities bought some. The beauty of this machine was its user-friendliness; it was not difficult to learn how to use it, and the chief requirement of the Paintbox artist was an artistic flair. The program could mimic virtually any hard-graphic technique: Tell the Paintbox that your touch-pen was an airbrush and it behaved like one; similarly, it could imitate a pencil, a brush, a crayon, nearly anything available to the conventional artist. Although Quantel remains in the forefront of the field, the Paintbox heralded a proliferation of very useful electronic graphic devices and its name has now attained the honor of generic status, along with Kleenex and Hoover.

New to the market in this area, and in postproduction suites around the world, is 3-D imaging and computer animation equipment. Contrary to popular belief, the computer does not do everything in these processes; the initial creativity is human, as are the parameters within which each computer construction is built. The computer is used to assimilate these parameters and fill in certain gaps (in the motion of an animated figure or the external appearance of a 3-D object, for example) and to speed the process along. It's still an artist's tool, albeit a highly sophisticated one.

Computer advances have also had an impact on the traditional art of editing. Although news programming and some entertainment programming is edited live—as it happens—most television product is edited in postproduction. There are sound reasons why we will always have editing suites. Much of what is shot is done on location or between locations and studios and cannot physically be done in real time. Scenes must be re-lit, actors must change costumes and rest between takes, and few scenes are done perfectly in the first take.

Most suites have computer controllers that allow remote operation of a number of playback sources. Typically, these are videotape decks feeding through the edit controller into a single recording deck. The format may be open reel, cassette, or even digital discs, but the mechanics are essentially the same. The editor can then play back any bits of the raw footage and record them on the edited master tape in whatever sequence they are required. Depending on the complexity of the system, this may be one playback source or many, and it may include any number of the fancy wipes, fades, flips, or gyrations available on the control-room switcher. Still, the function is the same as it is in the traditional film editing suite: splicing together disparate bits and pieces to make a coherent whole. It's just faster and a lot less messy!

Again, the trend has been to move the controls from an out-of-studio suite to the studio control room. Much field footage used in news reporting is shot on the Sony Betacam format (not related to the ill-fated Betamax) and is brought into the station in small cassettes. Each of these cassettes can be loaded into a Beta Carousel playback machine, which acts like a video jukebox and is controlled from the production desk in the studio. An operator in the control room simply sequences through the cassettes and cues each for playback by using a keypad. This eliminates the need to have a tape operator standing by during every newscast just to change tapes and play them back.

Digital Advances

It is now possible to edit video programs inside a computer without using most of the mechanical playback equipment required for conventional editing. These "soft" editing systems collect picture and sound information from conventional analog formats and store it digitally on discs. The editing process is then enacted in a somewhat conventional way until the finished piece has been assembled, at which point it is dumped back onto conventional video format for playback into a broadcast system. This would seem to beg the question, "Why not originate in a digital format, or, indeed, stay with such a format all the way through?" I can't think of a single reason. Can you? Apart from the fact that we've invested in a lot of expensive analog equipment and that cost will take some years to be amortized, there is no good reason to think that digital video will not be the way of the future: It is perfectly clean and crisp and does not degenerate in quality with successive re-recordings. You can edit a piece repeatedly, and it will still look as good as the raw footage you originally shot.

This is a big advantage. The various impressive special effects we see on film and video are the result of *layering*, or building up one image on top of another, until a wholly artificial world is created. The down side of this process is that the picture quality worsens with each layer or generation. With a digital format, we can have the best of both worlds: flawless picture quality and the ability to layer images to our hearts' content.

What All This Means to the Designers

Perhaps you begin to see where this is going. At one time, there were set designers (art directors), graphic designers, and animators, and rarely did they meet. With the increasing use of computer editing and animation and with the arrival of digital video, we can easily and routinely mix live images with graphic images or even 3-D animated graphic images. The graphics can actually *Be* the sets. Real actors can perform alongside animated cartoon characters. The possibilities are, as they say, limitless. The prospect of an electronically-generated news anchor such as Max Headroom is no longer a fantasy; many a news director's dream may soon become a reality!

Assembling the Product

Apart from the system, the process itself has a set of mechanics by which it functions; it may not be glamorous to think of television as a sort of machine, but it is not entirely inaccurate either.

The mechanics by which TV programming is put together may be thought of as systems of assembly. The progress is from idea to realization and is essentially the same as that of a manufacturing process. It is the creative staff—the writers, designers, and directors—who initially describe the idea and who implement the process of realization. This is done by means of various written documents, each of which is an attempt to define the idea in successively more specific terms.

For the writer, these documents will be numerous scripts, most of which are revisions of a preliminary first draft. In meetings with the other creatives, the writer (or writers) will expand on some things, eliminate or modify those scenes that may be too costly or technically unfeasible, and generally be forced to compromise the original idea to the realities of the medium and industry.

For the director, shooting scripts with shot, sound, and effects notations will evolve from the writer's scripts, and will also go through a series of revisions. The final documentation will be made in the finished shooting script and a set of storyboards, prepared in conjunction with the designer(s) and the director of photography, or chief camera operator.

The production designer will take the rough draft and develop a set of rough sketches depicting typical scenes from the script, usually after a preliminary scouting tour of possible locations. After a succession of meetings, these locations will be finalized and the number and scope of studio sets will be decided. The script may then be committed to storyboards and construction drawings and color references prepared. This planning period may take weeks or months; it is necessary to define the scope of the project and thus to estimate the facilities, crew, and budget required for completion.

This is, again, the way things are done at the high end of the market; most of us work at a somewhat less well-funded point on the scale and thus have fewer resources at hand. In a typical station situation, the process will likely be similar, though with less meat on its bones. For a news or talk show, for example, the role of the writer may be nonessential and that role may be assumed by a producer-director. Nonetheless, the need to define the idea behind the project is still vital and the roles of the creative team members in this process is the same. Too often, the creative brainstorming and the preproduction documentation is skipped over or dealt with too lightly and the end product suffers for it. A plain, dull setting flatly and brightly lit is almost always the result of too little time and attention devoted to the creative process and its implementation.

Once the idea has been finalized, agreed upon, and committed to paper, the process of realization begins. Talent is auditioned and hired; crew people (if not already on staff) are found; and the physical construction of the various elements of settings, costumes, props, and lighting is begun. As noted before, these functions may be done in-house or jobbed out to freelance suppliers and construction houses.

As this is going on, the production office staff is confirming availability of location sites, dates, and studios and plotting the production schedule. The production schedule often takes the form of a huge wall chart, typically with erasable marker notations or bits of colored card so that it remains dynamic and can be reconfigured as an ongoing part of the planning phase. Once everything is in place, the final shooting schedule may be printed and distributed to those who need it. (The term final as used here is a relative one; nothing *ever* goes exactly according to schedule in this business because there are simply too many variables, and Murphy's Law applies as always.) The shoot begins, and the scenes are shot in order of convenience, rather than in the order in which they appear in the script.

Exigencies of site availability, equipment availability, and many other factors dictate that shooting out of sequence is an economic necessity. During the shoot, the production designer becomes a resource person for visual continuity; since scenes are shot out of order, it is essential that everything look as it did in the previous days' shoots. The same pictures must be on the walls and in the same places; the lighting must be the same; the costumes and props must be as they were before, or in some way arranged according to the logic of the events as they will be presented in the final edited version (as written in the script). This is a tough, mind-boggling task, and requires an eye for detail. Once the shooting is done, there is rarely (if ever) time or money enough to go back and correct such continuity errors after they've been discovered in the editing room.

What happens if the program is not a once-off single project, but a continuing series or a daily talk show? All that changes is that the staff must do all of the above tasks constantly, on a daily basis, rather than doing one of them at a time in sequence. If you think the continuity responsibility as I described it here looks daunting, think about designing for a major drama series in which three or four episodes may be in various stages of production at any one time. Burnout, as you may think, is not uncommon at that level of production design; in fact, there are some designers now who are specialists in program pilots, doing only the origination of a series and perhaps the first season, and then moving on to another project. It's not a matter of having a short attention span but of preserving one's sanity.

At the editing stage, the production designer often bows out, perhaps occasionally dropping by to see how the edit is going. Editing is a tedious process, which requires great patience, and most of the designer's comments at this stage will seem unwelcome because they tend to begin with, "If only we'd shot that a bit differently..."

If I've succeeded in painting a picture of a somewhat cumbersome, rather complicated machine that we call video production, then I have done what I hoped to do. The process of creating television requires the talents and skills of a disparate group of people and relies on a complex system of highly sophisticated electronic hardware to realize an idea in light and sound. A visitor from a previous time might well think this a ludicrous endeavor. It is certainly an improbable one, and to get involved in it is to set yourself on a road full of pitfalls. Still, the reward of success is directly proportional to the likelihood of failure or no one would enter this profession. The degree to which you find success is closely tied to your understanding of the mechanics of the system and your ability to make them serve your ideas; approach it in this way and you may actually stay in the business long enough to retire.

10 Interpretation

Researching the Interpretation

When I was studying design, the notion of interpretation always came up in the context of drama or dance or opera. The script was always introduced as the first element of the interpretative process. With that as a starting point, the designer would research the styles that might be appropriate to the piece and would develop the design idea from there. By a process of research, study, and assimilation, the designer would develop a visual treatment for the production, which both served its needs and expressed something of the designer's own style.

This scenario is, of course, the ideal. It fails to take into account the many and varied real-world conditions that always come into play. Budget limitations, shortages of time and staff, and conflicting ideas of good design among the different creative personnel conspire to limit the scope for creativity.

Still, there are many cases in which the designer feels the satisfaction that comes from the achievement of a degree of self-expression. Often, this is somewhat unexpected and happens regardless of seemingly insurmountable limitations.

A beginning in theatre or television drama production is useful in this area because it gives the designer a more concrete starting point and more well-defined parameters within which to build a design solution. If the script calls for a realistic environment, the solution will be easier to reach (in most cases) than one that does not define the terms of the environment so specifically. Unfortunately, most television production (particularly that which is most accessible to the beginning designer in her first job) does not come with a script. The novice designer will usually be assigned a talk show or news show (or, perhaps, a Saturday morning kids' show) with only a page or two of description and a prospective script or outline for the pilot episode.

The paperwork on which a program proposal is based typically extends to a description of the format or mix of component parts that will make up each program segment. A careful analysis will reveal a consistent pattern to most continuing programs; the sequence of component parts usually follows the same order, and each segment is approximately the same length of time in each installment. A news program, for example, will follow its format religiously, devoting the same amount of time to each segment and presenting each type of story in the same order from one newscast to the next. (You may note that the total amount of *news* in a typical half-hour news program is around 11 minutes.)

If the program's developers have done their work well, the program brief will contain a fairly good profile of the prospective audience. Demographic profiles can often be drawn accurately, and can give the creative team its best clues as to the style of presentation that will work.

Typically, the program brief will be built around a number of people to be on camera (heads) and a description of their relationships to each other in the context of the program. A news program, for instance, will begin with either one or two anchors who are the principal news presenters. These are the people who open and close the show and who do most of the story introductions and hand off to the other on-camera people. Secondary heads might be the sports reporter; the weather reporter; and any of a number of various feature reporters doing film reviews, human-interest stories, consumer reports, and so on. This formula is standard for most news broadcasts and varies little worldwide. If you watch television with any regularity, you're already quite familiar with it.

Developing the Design Solution

Having established the categories of personnel to be accommodated by your studio set, you may now proceed to place them in a geometric pattern in relation to each other. This usually brings the designer smack up against one or more studio walls; there is rarely enough room to allow for a perfect accommodation for each on-camera position. Depending on the priorities that prevail, the news may be given a fairly substantial studio or it may be packed into a room that looks like it was once a broom closet. I recall an insidious construction called a *presentation studio*, which was about 10 feet square, and allowed for a single head behind a very small desk. It was intended for brief announcements during station identification breaks; however, it was eventually used for news presentations and even for a children's puppet show!

Figure 10.1 Schematic of a typical newsdesk setup

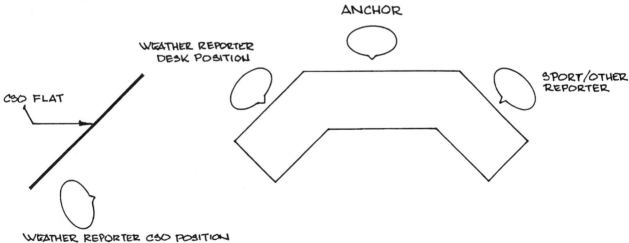

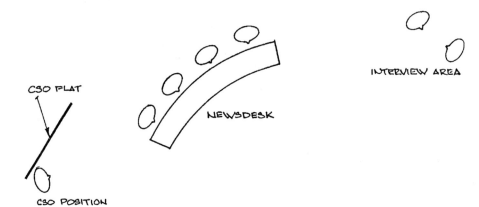

Figure 10.2 Schematic of news setup with CSO area

After the essential positions, such as the anchors and the chroma-key panel (for the weather map), are established it is possible to design ancillary areas for the special-interest presenters and for the occasional interview. The objective in designing these secondary areas is to make them all-purpose so that one head can leave quietly and be replaced by another while the shot is on one of the anchors. (This is another good reason for the anchors to do lead-ins for stories to be presented by the secondary talent: it gives them time to change seats.)

There are usually two forces competing for studio space: the set and the cameras. It is important that the designer be familiar with the cameras in use in a given studio, down to the exact dimensions of the pedestals on which they are mounted. It is critical that the camera operators be able to move very quickly from one point on the studio floor to virtually any other point. Thus, after allocating space to the talent, the designer defines traffic lanes for the cameras and their operators. The designer must understand the anticipated camera vocabulary (that is, the types of shots that will be necessary for the program to work). If the director is going to want over-shoulder shots to do interviews, the designer will have to allow for these both in terms of controlling what is in the shot and allowing room for the camera operators to position the cameras where they need to be and to travel quickly and easily between positions.

In addition, many studios also require such hardware as microphone booms, teleprompt equipment, and other paraphernalia all of which eats up valuable floor space. Most studios eventually look something like Granny's attic, because nothing is ever thrown out. ("It might be useful one day, you never know." It is a truism in the mythology of the studio that you always find a use for that old second-Empire fireplace, left over from a drama series 20 years ago, on the day after you throw it out.)

You may notice that we haven't done any interpreting of this news program yet. It is important that you identify your physical limitations before embarking on the creative and expressive phase of your design. It is also wise at this juncture to remind yourself that the director is often less interested in the aesthetics than in the traffic patterns, and you may get anything from blank disinterest to abuse when you fish for input that will help you with style choices. A good situation is one in which the director is at least able to read the floor plan and visualize it in three dimensions; typically, you will be required to build a model in order to communicate your design ideas to the director, and even then the director may not really fully understand what the set is going to look like when it's installed in the studio. I remember designing a children's show with a 2-foot-high platform, and having to build a scale model of a studio camera to convince the director that it would be possible to shoot *over* the platform. Those of us in the visual arts tend to assume the capacity to visualize; many people have no such capacity, and some of those people are directors.

Lateral Thinking

I can't stress too strongly the importance of knowing your audience in the matter of interpretation; finding a visual style within which to present a program must always be done in the context of the typical *viewer's* interpretation or you may fail to communicate what you want to. Frank Schneider, of WNBC in New York, tells of the designer who was called in from Miami to design a news set for New Yorkers. What he wanted to do was not bad design, it was just not the sort of thing that would appeal to a New York audience: The colors were too bright and the overall treatment too soft to command serious attention. Knowing what your audience is accustomed to, the color palette to which they relate every day and with which they feel comfortable, is critical to putting them at ease and telling them how to interpret the content of the program.

From having spent his life in New York, Frank knew that the treatment of the news set should be reserved and rather serious, though with a

Figure 10.3 WNBC-TV news set with designer sitting in. Design: Frank Schneider

crisp, hard-edged texture. Anyone who has lived in New York knows that it is a fast, hard-edged society and most of its people will identify with visuals such as this. In the end, the set treatment was done in an indigo blue fabric stretched over square panels, the desks done in grey linoleum with a gold metallic stripe across the front edge. The set is quite deep front-to-back, with canister down-lights built into the headers, and features a bank of monitors at the back of the principal anchor shot.

The effect of all this is a clean, high-tech look that is associated with up-to-date communication technology and gives the impression of an electronic nerve center crossed with a modern office. This is not, of course, the actual newsroom, and some of those monitors at the back are dummys with static color transparencies in them, which are lit from behind. (Frank tries to avoid having moving images in the background of the set at head level because it tends to distract from the talent.) Nevertheless, it has the look of a highly sophisticated news center and thus buys for the program an extra measure of credibility.

Inherent Interpretation

It is important to remember that the substance of television is illusion. This works to the advantage of anyone with a creative mind. Just as on stage, the actual depiction of reality on television is something of an impossibility. It is still the subject of debate among media theorists, but anyone who has been in the field with a news crew knows that the entrance of a television crew into a real-life situation immediately changes it. I recall hearing about a film student in New York who planned to shoot a street scene on the Lower East Side of Manhattan, in an immigrant community, and found that when her camera was set up and ready to roll there was no one left in sight. The camera is an intruder on reality, and *must* alter it in some way simply by its presence.

The camera lens is selective; it can't see as much as the eye, move as quickly, or adjust its focus as quickly. It sees what its operator wants it to see and in the way that the operator wants it to see. Everything you put on camera is interpreted (whether you intend it or not) and every picture you put on television makes a statement. It is thus important that you understand the medium as a tool of interpretation, and use it to communicate what you want it to.

Tools at Hand

As a designer, your tools of interpretation are a bit more subtle and subliminal than those of some others on the creative team. The writer, obviously, has the most direct means of making a statement and the performers and director the next most direct. To the extent that the designer has input into the camera scripting, he or she will increase the capacity for interpretation.

Much as in stage design, the greatest interpretative task in television design is the creation of an environment. The set establishes a visual context for the events that occur within it and influences the meaning that is derived from them. This task involves the physical arrangement of the set elements (and, consequently, of the people in shot) and the color, texture, and significance of everything seen in the background.

Getting It Right

Recently, I saw a setting for a talk show, produced by a major-market station in an Eastern city. The production brief called for a casual living-room set in which the members of the discussion group (and, presumably, the audience—though no mention was made of them) could forget that they were in a TV studio and just talk freely. First, let's acknowledge that this is an impossible task: The sheer weight of equipment hanging from the lighting grid, the various crew in attendance, and the intense heat of thousands of watts of light make it impossible for anyone, no matter how obtuse, to forget that he's in a TV studio! This designer came up with a perfectly symmetrical box set: three sofas (one upstage center and two facing each other across the set) and an oriental rug in front of a row of rather unconvincing French doors. The set was lit like high noon and was shot straight up the middle: as boring and static a composition as one could imagine. There was no dressing: no props here and there to suggest that it might have been a room that was lived in; nothing to add texture and interest to the background of the single head shots. In short, the designer missed every opportunity to interpret the intent of the program in the visuals.

The program brief had called for a set done in soft pastels and generally warm colors to put the guests at ease. What the designer did wrong was to regard himself as just another functionary; the set descriptions garnered from program briefs and directors should properly be regarded as road signs leading toward a final visual solution that is more specific and much more a result of combined input. The difficulty of this method is that it requires the designer to *think*. The capacity to use one's brain and the personal fortitude to spend hours at the drawing board and even more hours in the prop shop or the Salvation Army finding dressing pieces for the set can often be the difference between a dry, boring, generic set that says absolutely nothing about the show and one that works (or even works well).

What sort of people are going to watch the show? What sort of personality does the show's host project? What kind of subject matter is likely to form the mainstay of the program's content? Is the bulk of the program going to be done in studio or will most of it be footage shot in the field? Is the slant of the show toward its subject matter serious or is it light? How serious? How light?

These are some of the questions the designer should ask at the beginning of the project. Each question can be answered in somewhat general terms at the initial preproduction meetings; however, the thorough designer doesn't usually settle for the first answer. Each of the questions that will come to mind in the development process can be pursued further; there is always another level of detail to which each line of investigation can be pursued, and the process is often not truly at its end until the show is off the air.

The mental process is something like the construction of a graph: along one axis is the list of questions about the shape the program is supposed to take (what it's supposed to do and say) and on the other axis is the list of various visual treatments that might aid and abet in the successful accomplishment of the program's objectives. If you were to make such a chart, it would probably be quite large; it may help you to reach a truly satisfying solution and for that it would be well worth the trouble. Every designer I know has a different set of mental tools to help with this process, and you will need to develop your own. It is important to treat the design process as one of evolution; it is also important that your design ideas are open to change and growth throughout the preproduction and production stages in order that the final design solution is synchronized with the program's objectives. To that end, the use of a set model together with a number of sketches of likely camera shots can be very helpful, provided that the designer is not in love with them and remains willing to throw out an idea which doesn't work no matter how nice the sketch or model is.

Ongoing Input

The fact that, unlike theatre set design, television doesn't dispense with the designer's services once the show has opened is good for employment prospects. As a matter of routine, the set needs to be reconstructed every time the program is taped (or aired, if it's live). The flats and other set pieces have to be pulled from storage and re-erected in the studio, the finish surfaces touched up where they've been marred, and the various items of set dressing placed in their rightful positions. All of this is usually done under the supervision of the designer, and her expertise will usually be called on by the lighting crew to make sure that they have done their work well and consistently with established program continuity.

Although in theory the identifying elements of a program series make up its signature and do not change once they have been established, in reality a program is in a continual state of evolution even after it has gone on air. This means that the involvement of the designer in servicing a show requires more than merely duplicating the established look week after week. As the program evolves and grows, the director and the producer may identify ways in which certain elements of the production may need to be changed in order to anticipate the changing expectations of the audience, personnel changes, or the natural evolution through which the show must progress. The better designers will not only be prepared to respond to alterations and changes but also will continually be alert for things that can be done visually to enhance the program and facilitate its growth. This usually comes down to ways in which the program signature can be supported and the format made more effective.

Some Good Examples

The work of Derek McLane for a program called "Real Life," featuring Jane Pauley on the NBC network is a fine example of an appropriate, simple yet elegant design solution. The program was a magazine–documentary format show, emphasizing human-interest subject matter, but sometimes delving into harder news issues. The bulk of the program was field footage, the studio set being used only for *top-and-tail* commentary by the host. The set, then, was extremely simple: just a corner of a room in a city apartment or loft, a window to the right of the shot, and a chair on which the host sits. The walls were light grey and textured, so that the strongly colored light coming up through the window from below gave the set its image and the show its signature. There was exactly as much in the shot as functionally necessary and as necessary to give a sense of atmosphere without looking over-designed or over-dressed. Soaps are similar when they're designed and lit well: there is a sparseness, an absence of detail, that is consistent with the larger-than-life (and sometimes downright incredible) plot lines and characterizations that are the trademark of these shows.

In the best working situations, a program will have a creative team (director, production designer, lighting director, and graphic designer) assigned to it for a long stretch of time. This ensures a continuing relationship among these people, that they are in constant contact with the program, and that they can stay in touch with the subtle ongoing changes that direct the program's evolution. In many cases, however, this is not the situation. As with the feature series produced on film, often the locally produced show will have an origination team that is different from the team that picks up the show for the rest of its run. Especially with lighting and sound personnel, there is often a daily or weekly rotation schedule, which means that the people responsible for the look and sound of the program may be different from one day to the next.

The production designer is typically the one member of the creative personnel who is assigned over the long haul, and it is frequently she who must try to maintain the program's signature look from one episode to the next. If the relationship between the designer and the lighting staff is a good one, this may not be difficult; if not, there can be a great deal of friction leading, all too often, to a poorer result on camera. This is one reason why it is essential that the production designer have a solid understanding of both the aesthetics and the pragmatics of television lighting. If you, as the designer, can convince the lighting director that you know what you're talking about when the two of you discuss the look you want for your set and can describe the means you think will best achieve that look, there's a very good probability that you'll get what you want. Refer to the photographs of the work of Frank Schneider (Plate 5) and Alan Farquharson (Plate 2); this sort of success in achieving of a specific look is only reached through careful cooperation between the lighting crew and production designers, and reflects a high level of mutual understanding.

Interpretation is both a creative and a mechanical process. Initially, you'll be required to make visual interpretations of ideas that will be expressed to you in general terms of mood, personality, or image by people whose facility is verbal rather than visual. In most cases, they will not speak your language; the process by which the final look of a given show will be resolved will be one of presentation and modification, the director looking at roughs of your ideas and responding in ways such as, "it's not

---

Top and tail refers to a presenter, introducing (*top*) and wrapping up (*tail*) and edited story, often from a studio set.

quite as _____ as I'd imagined—why don't you...?" Many directors can only tell you that it isn't what they had in mind and hope that the preproduction process will take the form of a gradually narrowing spiral, until one morning you will show them something which clicks with their own preconceptions. The benefit of such a process is that it allows the designer to ease the director's thinking in another direction, and usually results in a product that is more truly collaborative than one in which the director simply uses the designer as a functionary to express her own preconception.

Distortion and Fantasy   11

Degrees of Interpretation

I've stated that it is impossible to accurately represent reality on television. The fact that it is being processed through the system and the change of context from real life to the screen of a television set will ensure that what is being presented is not reality itself but rather someone's interpretation of it. The same is true of film or theatre; the context alters the meaning to some degree.

We may, therefore, regard distortion as inherent in the medium and the difference between the television depiction of realism and fantasy to be a matter of degree. This is an area in which television has traditionally gotten into trouble: The inability of some viewers to separate what they see on the box from real life has brought blame for some types of antisocial behavior down on the heads of TV people. To be fair, such associations were also made with film and theatre in their time and are, therefore, nothing new.

People making TV programs are generally not unaware of the interpretative nature of the medium and have rarely let fear that someone, somewhere, might commit a crime based on something seen on television stop them from indulging their creative fancy. The primary directive of the medium is to entertain and, at its best, to stimulate the imagination of the audience. As with other art forms, it works best when it stimulates the minds of the viewers and when it poses questions rather than preaching about the answers.

The spin put on a story by the way in which it is presented can vary anywhere from a subtle hint that one particular character is the stronger one in a given situation to the wholesale alteration of the nature of the universe in which the two characters must function. The more blatant the removal from the appearance of everyday reality, the more apparent the contribution of the design staff, and the more likely that they will earn an Emmy for their efforts.

The creation of any environment other than today's reality (whatever that might be) involves a substantial effort from the design people, and a substantial element of interpretation and even fantasy. In many cases, the literal, accurate presentation of the everyday world of the past would leave it inaccessible to much of the modern audience; even the production of a historical drama demands that the look (and even the language) of the time be extensively interpreted to make it palatable.

The critical element in the effective use of interpretation or fantasy is remaining true to the essential theme of the work. *Star Wars* was a landmark event in the field of film fantasy, but what held the attention of the audience and ultimately provided the core upon which to build the film was the classic plot of the adventure/romance. Lucas and company never lost sight of that, and the tremendous success of the *Star Wars* films was due in large part to their ability to make the flights of fantasy subordinate to the elements of human emotion with which all of us identify, even if they're being expressed by R2D2.

The best scripts have a strong theme, one that comes through because of its characters and their interaction and because of the context in which it is presented. The production designer has a great many tools of interpretation and distortion at his disposal; the *intelligent* designer knows how to use them to support the theme of the work without subverting it. This is a skill both of perception and application that is not usually easy to acquire, and typically comes with maturity and experience.

Classic among the works of visual fantasy is Ridley Scott's *Blade Runner*. This film entailed the creation of a very complicated perversion of today's society, in which nearly everything resulting from the technological advancements of the 20th century had gone wrong. The extent of visual detail and the unrelentingly dismal atmosphere made the set design a significant player in the story. Still, the touch of the designer was not so heavy handed as to drown the actors' performances or diminish the intensity of the human conflicts central to the theme of the piece.

Similarly, the Public Broadcasting Service (PBS) produced a television version of Ursula LeGuin's novel, *The Lathe of Heaven* in which the central character woke up to a different reality every day. The theme was the relativity of human perception, and the video trickery available in the late 1970s was subtly and intelligently used to present a story of the triumph of love over reality.

As with all matters of style and touch, the best tutelage is to study the work of those who have mastered the use of the available tools. This includes the work of the masters of film fantasy; the advance of video technology has put the full catalog of effects once exclusive to film at the disposal of video creatives, and frequently at a substantially lower cost.

The Mechanics of Fantasy

We define realism in a number of ways; generally, these can be broken down into two categories—physical and temporal. If a thing looks like things we are accustomed to seeing every day, we're likely to accept the appearance of it on-screen as realistic. If a sequence of events happens in the order in which we think they might really happen, we tend to accept those events as being realistically depicted. If these parameters are adhered to, we are inclined to accept the whole package as a realistic portrayal. These parameters are not too strictly drawn, and have been stretched over the years by increasingly imaginative use of fantasy devices in the film and television media and the consequent willingness of audiences to accept them. From "My Favorite Martian" to "Alf" and "Alien Nation," the con-

Figure 11.1 A shot from *The Singing Detective* with Michael Gambon. Courtesy of BBC/Lionheart Television

cept of extraterrestrial beings situated among everyday Earthlings has come to be easily accepted and moved from a simple comedic gag in the first instance to a source of more serious drama in the last. The adherence of the program to a normal depiction of reality and a normal sequence of events made the acceptance of alien beings as serious dramatic subjects an attainable goal of the series' writers.

More difficult to accept in the context of reality is the disruption of time reference or the juxtaposition of unlikely elements in ways that we are not accustomed to seeing. Both as a novel and as a video drama, *The Lathe of Heaven* was disruptive and difficult to follow. In nearly every chapter of the book, the protagonist awoke to find a new reality with a new set of rules and assumptions. This was quite difficult to depict and more upsetting to the viewer trying to follow the plot than the idea of a space creature living in the house next door. Dennis Potter's "The Singing Detective" was an equally disturbing piece simply because he took liberties with time, place, and character relationships. Within the context of the present-day or within the context of the traditional detective novel, there was nothing particularly out of order; the juxtaposition came with the frequency and abruptness with which the story jumped between the two realities and the spillover of characters between them.

There are a number of conventions of time and space that will cue the audience as to the context within which it is to interpret certain events. A short *dip to black* or the insertion of a graphic indicating a change of time or place will be universally accepted as such, and permit the action to jump days, weeks, years, or thousands of miles without disorienting the audi-

---

Dip to black is a short fade from a picture to a black screen and back to another picture.

ence. A *whip-pan* can have the same effect. A scene shot in *soft focus* will automatically be taken to indicate a dream sequence. Such conventions must be understood and adhered to; otherwise, the audience may lose interest or mistake a serious piece to become. David Lynch's *Wild At Heart* had many mixed cues for the audience and left many a bewildered patron in the cinema in which I saw it; few knew when to laugh or cry and no one knew how to interpret the appearance of the Good Witch from *The Wizard of Oz* at the end!

As much as we stress our individuality, most of us have a shared routine in our daily existences, which serves as a common basis of experience upon which to make assessments of the things we see. The net effect of the so-called global village has been the homogenization of world culture: you can get a McDonald's burger virtually anywhere on the planet now. People in the eastern United States and in Europe are speaking English like California surfers thanks to Bart Simpson and other products of the American film and television industry. I recall my disappointment at my first trip to England as I found it substantially less different than I had anticipated; "Dallas" was the most popular TV series and London was beginning to show the proliferation of American and "American-style" fast food emporia, which has now reached a point of saturation.

It is safe to say that this is the trend—over the years, the texture of our daily existence has come to be similar to that of much of the rest of the world and consequently, the language of the familiar and the fantastic has also become universal. This is good for the makers of film and television programs; it means that there is a larger market for their product outside the primary market for which it was intended. Just as cars are designed for the world market, it is now possible to create television shows for a world market, with the knowledge that the more subtle conventions of audience cues and contextual reference will largely be understood worldwide.

The disruption of the texture of routine existence has a shock value and a fantasy value that have long been understood and used to advantage. In its most elemental form, this is easy to achieve without special technical devices: For "The Incredible Hulk" it was only necessary to substitute actors, apply some green makeup, and cover the change with some quick cuts between tight close-ups. In the film *Flatliners* the effect was achieved by some very clever and subtle lighting changes and superb camera work. All that was necessary was the removal of the action from the context of reality by inserting of a giant green man or an unnaturally saturated lighting effect. In both cases, the disruption of the picture of routine daily life by the introduction of a single element was sufficient to cue the audience that they'd crossed the line out of "realism."

Trickier, but with equally strong dramatic potential, is temporal disruption. The *flashback* is an old convention, and sufficiently well accepted that few viewers of television would fail to understand its cues and follow

---

Whip-pan is a jarringly rapid lateral motion of the camera, which is usually a mistake.

Soft focus is achieved with a special filter over the lens, which makes the margins of the picture go fuzzy, but leaves the center sharp.

the action through many time changes. The vocabulary of flashbacks is old enough that we now have many cues available to indicate it. The simplest method is the dip to black, although the brief fade into soft focus, the rippling image, and the (now obsolete) flipping calendar or newspaper pages are well-established. Audiences are well enough accustomed to rapid time changes, whether backward or forward, that they will readily pick up on a sudden change of lighting or season, or the dress of the actors, and rarely find this an obstruction to the story line. One of the classics of fantasy was *It's A Wonderful Life*. The only specific indicator of the change from fantasy to reality was the presence or absence of the falling snow. Subtle though this was, Frank Capra knew his visual vocabulary and knew that it would be enough.

The alteration of physical reality is usually a more obvious tool of fantasy and usually also the one requiring more expensive tools and processes. In its simplest form, this requires only the creation of a wholly unrealistic environment in which to shoot the action. Most of the science fiction programs fall into such a category; classic series such as "Star Trek" and "Doctor Who" have relied largely on the creation of wholly fantastic sets to frame their fantasy plots. The use of a police phone box for a spaceship and a robotic dog gave the viewers of "Doctor Who" the double cue that the program was both a fantasy and a wry parody of itself and its genre.

The ultimate tool in the creation of fantasy is, of course, animation. While the use of unrealistic sets and actors in futuristic costumes buys a great degree of freedom from the constraints of realistic expectation, the animator's art begins with a blank slate and creates its own world—inhabitants included—without the constraints of any aspect of physical reality, including gravity. From the early days of animation, its practitioners took advantage of this fact and made the defiance of the laws of physics a staple expectation of cartoon audiences. This has, of course, been surpassed with the successful melding of live action and animation in *Who Framed Roger Rabbit. The* shock of juxtaposition has never been used better than in the opening sequence of that film, when the camera pans from the cartoon "set" to the real-life studio set. The media press made a great deal of noise about the breakthrough in computer animation, which the film represented, missing entirely the point which was most important—the creative insight of the film's artistic personnel and the seamless manner in which they constructed a fantasy reality and adhered to its rules.

The Machinery

The simplest mechanisms available for the creation of alternate realities or fantastic effects are the old ones. Use of unnatural lighting angles and colors can go a long way toward creating unreal effects, as can creative use of camera angles and distortion caused by certain lenses.

Polarizing filters, graduated color filters, and various other things can be stuck on the front of the camera lens to create different effects. (The classic soft-focus effect traditionally used to indicate a dream sequence was often done by covering the lens with a piece of glass and smearing that with petroleum jelly.) The standard lens in every musical or variety show's repertoire

Figure 11.2 Star filter shot

is the star filter, which takes any specular reflection in the shot and makes it appear to have spiked rays projecting from it, like the idealized (and incorrect) perception of the appearance of stars in the sky.

There is a large catalog of lenses and filters available, but the real workhorses of unrealism in video are behind the cameras. Because we convert visual images to electrical impulses in the process of creating television shows, we buy ourselves the ability to modify those images. The flow of electrons through the system is analogous to a river, and the technical term for a point further along in the system is *downstream*. Downstream from the camera can be any of a number of effects devices that can alter different aspects of the picture electronically.

The most fundamental of these is the time-base corrector, the purpose of which is really to synchronize different picture sources. It does this by grabbing single frames from several sources and spitting them out in synchronization with each other. It is thus possible to grab a single frame from an action sequence, or to alter the speed of an action sequence. It is also possible to "drop" frames from such an action sequence, giving a jerky quality to that action without altering the actual time from inception of the action to completion. This is the staple of much of Music Television's (MTV's) product.

Some of the simpler effects devices incorporate these techniques and extend to effects such as *pixelization*, which alters the texture of the image and enables the operator to break the image into mosaic patterns of varying size. These devices usually also enable their users to alter the overall color range of a scene and change the contrast between lights and darks. There are many such machines on the market now, and the range of effects available is diverse.

One of these machines, the Bosch DaVinci, allows us to identify certain colors in a scene and alter only those colors, leaving the other colors intact. Do you remember the Cherry 7-Up commercials, in which we saw one or two pink objects in a scene which was otherwise black and white? That effect was achieved on a system such as this; the trick is to identify a single strong color and make sure that it appears in shot only where it is needed. (It works better if the actual color used is not the color of the object in the finished effect.) For a commercial such as this one, we might paint the 7-Up can yellow (being certain that it is the only yellow object in shot) and then adjust the machine to remove the color from the scene and to insert pink wherever it sees yellow. This is a very expensive device for the achievement of a single effect; however, it wasn't invented to make this commercial. This system was intended for a much more pragmatic use: It was developed and is used every day to match the color of film and video footage, so that scenes can be edited together without looking like a direct color mismatch. Light, as you may know, is not white; it is, in fact, many shades of amber or blue, depending on the source. Often, footage shot under different types of light or at different times of day may not match the color of other footage. Sometimes scenes such as this must be edited together with other scenes to carry the story line along, in spite of a direct color mismatch. Devices such as the DaVinci can adjust the color balance in a number of subtle ways until a reasonable color match is obtained. The spin-off of this process was a novel visual gimmick that was quite useful in producing fantasy effects for several advertising campaigns.(Often such spin-offs are developed by technicians playing with equipment in spare moments.)Recall the comment earlier in this book about the advantage held by those who are accustomed to finding unconventional uses for things; whatever the technology, creativity and ingenuity are always salable commodities.

Nifty as some of this gadgetry is, there is a fundamental image alteration that has opened up the distortion and fantasy options in almost limitless ways. This is the old technique you will remember from our news setting, the chroma-key effect. When chroma-key became a viable tool, directors and designers went a little crazy with it. This was not necessarily a good thing because the early key systems were rather crude and far from seamless.

The system, you'll recall, substitutes another image wherever it sees a certain shade of a certain color in the master shot, and thereby presents us with the option of juxtaposing unlikely things in one scene or putting people in impossible places. With such a system, we can make people fly or transport them to places they've never been. The trouble with the effect is that it often alters the images slightly; the edge between the foreground person or object and the background image is often too hard, or seems to have a halo. This occurs because the keyer must be adjusted to clip correctly—to decide at what point to make the substitution of one image for the other. The effect of this has typically been much like a photo pasteup: soft edges, such as a person's hair, tend to come out hard edged and look fake. Also, some keyers are unable to insert shadows from the foreground image onto the background image, and making the foreground image look a bit two-dimensional.

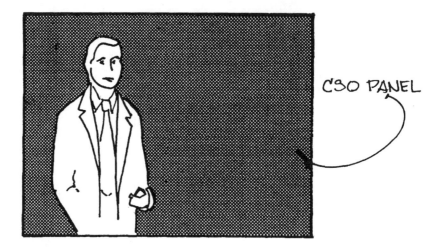
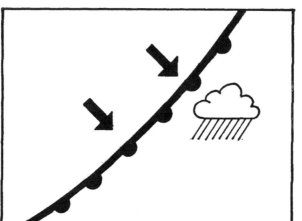

Figure 11.3 How CSO (chroma-key) insertion is done

There are now much more sophisticated (and much more expensive) chroma-key devices available that can superimpose one image on top of another seamlessly and with shadows. The major weather reporters (usually at the network level) typically have these now, and regularly cast shadows onto the maps and charts that are not behind them in the studio.

While this is a great improvement on the effects systems that only allow us to use one image source at a time, it is still effectively limited to a maximum of two or three levels of piling on. There is an effect in all analog video recording formats known as *degradation*. Each time an image is re-recorded, the quality of the picture deteriorates. Thus, each time you use an effect such as keying to pile on images, the quality of the final picture worsens until it becomes unusable. For this reason chroma-key is still most commonly used only in the news studio.

There are two types of key effects, however. Older than chroma-key but nonetheless useful is an effect known as a *luminance key*. This works on the principle that the brighter image (the one with greater luminance, hence the name) superimposes itself over a less bright image. For this to work, it is necessary that the images in the background be darker than the foreground keyed image. The advantage of this technique is that it generally gives a cleaner picture, with few of the undesireable artifacts that often spoil a chroma-key construction. In fact, the special effects used in the fantasy series "The Flash" were created on digital video (and converted to film for distribution) using a luminance key effect to insert the figure of the Flash himself and the red streak that follows him around. The streak is created and inserted separately, and the total effect of speed is a very clever use of stop-frame animation, put on film in the traditional way.

Piling-on is still an option to those in the fantasy-creation business. However, we are now in the early stages of the age of digital video; the technological advance that will revolutionize video is here. There have been many attempts to improve image quality in the analog formats, and some of these have been quite successful. Nonetheless, the inherent ability of digital recording formats to retain image quality over many levels of re-recording leaves analog video in its dust. This is the most useful tool yet developed for image alteration in the television industry, since it allows all of the layering, stop-frame, color and texture manipulation of film or analog video formats without the degradation indigenous (in varying degrees) in both.

Animation

The ultimate fantasy medium is, naturally, animation. The animator begins with a clean slate (a blank frame, actually) and builds everything that is needed for the scene, including the performers. The beauty of this is the ability to wantonly disregard the laws of physics, time, and space for the sake of the story. Animation was once regarded exclusively as a comic medium; in recent years, though, some filmmakers have ventured into serious animation work.

Traditionally, animation has been more a film tool than a video one. The nature of the film-exposure process has made this a natural association:

Since a frame of film is exposed individually, it is quite easy to do stop-motion animation. In this system, one or several frames may be exposed, the scene altered to show a slight advance in the action, and then another frame (or set of frames) exposed. If you think this a very long, tedious process, you're right. Animation requires a very high level of patience and precision, and is still a labor-intensive process. While the style of an animated series such as "The Simpsons" is established by its art director (located in Los Angeles), the actual frame-by-frame artwork and camerawork is executed in Korea, where labor costs are much lower. Because of its easy compatibility, film is still the medium of choice for animation.

Analog videotape is not easily able to record a single frame. In normal use, it takes about 5 seconds for the tape to get up to recording speed in the average broadcast video recorder. Imagine the waste of time and tape stock if an animator had to allow 5 seconds of roll-up time to record one or two frames of animation. This would be far too cumbersome, both in the shooting and in the editing, to be feasible. This has been the way of things until the advent of digital video.

With the coming of the wave of digital video graphic devices came the ability to generate pictures for frame-by-frame animation, together with a means of recording them that was as easily adaptable to this purpose as was film. Once these video graphic machines were established in the art departments of television production facilities, their manufacturers developed and sold upgrades of various sorts, including some with the ability to construct images in three dimensions and to create apparent motion (animation).

The computer that drives such a system must have a very large memory capacity, and cost is prohibitive to anyone other than a commercial video production house, large station, or network facility. Having said that, computer animation as it is done on the top-of-the-line hardware is generally more efficient than film animation, because it eliminates the extra step of carefully photographing the artwork frame by frame.

When such computer-driven systems are applied to the art of animation, they do the same things they do for still artwork: They simplify the task and speed the flow of work. Animation consists of several layers of artwork, and thus is the same as piling on images for conventional television applications (such as news broadcasts). There is typically a background with several levels of animated figures laid over it. The background does not generally change; the camera may pan across it, but the artwork itself remains static. The animated layers—traditionally done on *cels* of clear plastic—are laid over the background art, and consist of a series of still pictures, each of which represents a small fragment of a complete action sequence. In film animation, each cel is laid in place, shot, and removed to be replaced by the next one in the sequence. This represents a great deal of art and weeks or months of painstaking work, depending on the length of the piece.

Computer graphics do not generate ideas. They merely speed the process by which the artist (who does generate the ideas) renders her ideas and facilitates the process of revising and altering those pictures. In addition, computer graphics produce a first-generation source of artwork because they eliminate the need to put a hard graphic in front of a camera and shoot it. Apart from saving a step in the process, and thus saving time, this also eliminates one step of image degradation.

Computer animation does not eliminate the need to workout each step in an animated sequence, although it does eliminate the need for an individual artist to produce each step. The computer animator can construct a sequence of actions by defining the beginning and end points and drawing a certain number of points in between. The computer program can then fill in the required visual information to make a smoothly flowing action sequence.

The computer can also simulate a camera's movement. The background artwork will typically be generated within the computer itself (although it can be shot on a video camera from hard copy and inserted into the system). The system can then track across the background at the rate instructed to and insert moving foreground figures on top of that.

That's not all (folks). The really sophisticated machinery can think in three dimensions. A real 3-D object can be scanned by a camera and fed into the system or (much more difficult) a 3-D object can be created in the mind of the artist and its shape fed in. This can be a very lengthy process, but, once completed, it permits the system to manipulate these images in any way imaginable.

Having attained these mechanical functions, the developers of these apparatuses pressed on, improving resolution and detail capabilities, enabling them to mimic the images of reality. It is still, however, the mind of the artist at the helm of the system that decides what images to use and how to relate them to each other and which gives its signature to the piece. As with any computer software, each system has its own vocabulary and its own logic. Each requires a certain amount of training for the operator to use it and a good deal of experience to master it. In the same way that someone wishing to learn to paint with watercolor must spend a long time learning the idiosyncracies of the medium, so must an aspiring computer

Figure 11.4 Graphic design in the computer era. Cory Resh building an animated sequence at Modern Video Productions, Philadelphia

animator learn a great deal to make that a tool for his artistic expressions. In a static medium such as print, it is difficult to convey the dynamics and excitement to be found in computer-driven art and animation work; the best I can do here is to encourage you to weasel your way into such a studio and beg to be allowed to sit in a corner and watch for a day or two. You'll come away marveling at the capacity of the machinery and the patience of its operators and artists.

In typical television practice, such equipment is most commonly used in news production. Some years ago, I was in the paint box facility at the BBC's Lime Grove facility in London. There had been an airplane accident recently, involving an airliner catching fire on a runway. Thumbing through stock photos, the artist found the plane in question. She put it under a *rostrum camera* and inserted it into the system. Unfortunately, the airline logo and color scheme were incorrect, so she input the logo from a travel brochure and reduced it to scale and overlaid it on the picture of the airplane. Then she altered the colors using a touch pen and added wheels (since the plane in the photograph was showing flight, with its wheels up) by taking them from another photo. She overlaid a cutaway view of the interior of the plane, which she created herself on the system, showing details of the location of the fire and the passenger exit points. Having stored all these picture components, she then built an animation sequence that showed the plane, then overlaid the cutaway view, then inserted arrows and labels indicating the fire point and the exits. All of this took perhaps an hour to complete. This is the bread-and-butter application of such equipment and the sort of situation in which most people working in this end of production design will find their first jobs.

At the other end, though, digital computer-driven graphic art and animation systems virtually eliminate the barriers to image manipulation and fantasy creation. We can *grab* images from the real world and insert them into our fantasy environment, or we can create images of fantasy beings from our own imaginations and animate them in three dimensions and thus make them interact with images of the real world. We can, of course, take images of real beings from one real environment and insert them into another, making them change size, shape, or color or making them appear and disappear, or fly through the air. When I stated that you can create practically anything you can imagine (at least on screen!), I did mean it literally.

Rostrum camera is a video camera mounted on a stand to permit artwork to be placed underneath it and lit, and thus converted to a video image.

Script and Storyboard

12

Building on the Verbal Foundation

Despite all the advances in means of visual expression, most often the germ of an idea (no matter how visually oriented) is verbal. Whether a television program is a fully scripted drama or simply a talk show, the first incarnation is likely to be done in a written format. At its most basic, a program proposal may be written in paragraph or outline form in its germinal stage. By the time it is accepted for production and finds its way to the first production meeting, it is usually in script form, with notations about action and visual effects along with the dialogue.

No matter how much the designer may feel at ease with visual expression, she must still be fluent in written language. Those who develop ideas for television programs are usually verbally biased and not often good at visualization. The task of the designer is often to make the bridge from verbal expression to visual realization, and the ability to draw visual clues from words is absolutely essential to this process. It is not enough simply to know how to render an idea or even to see it through to construction and installation in the studio: Virtually always the overriding need is to pick up on verbal cues and use them to spark visual responses and to further refine and interpret visual expression. You may present a rendering of your design idea to a producer and have it rejected as "not quite what I had in mind." Going back to the drawing board again and again until it looks something like what the producer had in mind will work—eventually. How long it will take may well be another matter.

If you have a good command of the language and you are able to express yourself verbally with a fluency approaching that of your visual expression, you may condense this process and come up with a much better product as well. I don't know how many times I've seen a designer present a first effort toward a specific design solution and, when told that it "isn't quite right," simply stand there mutely looking crestfallen before going home to sulk. A cleverly worded and poetic defense of your design may not guarantee acceptance the first time out, but it may be that the capacity to discuss design ideas on a higher level will improve the flow of information and result in better communication and a better collaboration. It may also be that a person who crafts words will be less suspicious of someone who is also fluent with words and may be more inclined to like what you present. I'm not saying that I always pitch an idea with flights of lyrical

Figure 12.1 A storyboard

poetics, but I have known such pitches to make that little bit of difference that pushed the process over the top!

When you begin the development of the initial set of *storyboards*, you'd better be comfortable with verbal imagery and with your ability to sketch. The one element of the rendering process that inevitably goes through many stages of evolution (and thus is reworked most) is the storyboard. If you're fortunate (and you play it correctly) most of the "It is not quite what I had in mind" reactions will come during the initial stages of storyboarding rather than later, when you've spent several days or weeks developing a meticulously detailed model.

There are two kinds of scripts which you're likely to encounter: The writer's script and the directors' (shooting) script. Depending on the stage at which the designer is called in, it may be either. The writer's script contains less visual information, making reference only to the most general shot information such as cuts, fades, dissolves, or special effects. The shooting script is very specific and represents a much more advanced stage in the evolution of the idea. This is the more useful script for the designers because it gives very specific shot information, and thus indicates more explicitly what is in each shot.

Typically, the shooting script is divided into two vertical columns. The left-hand column contains information about each shot, in the greatest detail possible, while the right-hand column contains the text of the dialogue, together with movement notations for the actors or talent.

There is a complete vocabulary of shorthand notation for use in the left column, all of which the designer must know and comprehend. If there

```
                    RESTAURANT SCENE


SHOT 1              EMPTY TABLE AT A SMALL CITY
WS CAM 2            RESTAURANT.  MAN AND WOMAN ENTER CAM.
                    LEFT.

SHOT 2       MAN    (HOLDING CHAIR) Our Usual.
MS CAM 3

SHOT 3       WOMAN  (SITTING)  Thank you.
MS CAM 1

SHOT 4       MAN    Want something from the bar?
MCU MAN
CAM  3       WOMAN  A V and T, please.  Rocks?

             MAN    I'll have J&B.  (LOOKS FOR
                    WAITER)

             WOMAN  So?

             MAN    Know what you want?  To eat?

SHOT 5       WOMAN  Yeah.  Sure.  (PAUSE)  I guess
MCU WOMAN           I'm not really very hungry...
CAM  1              I don't know, why don't you
                    order?

SHOT 6       MAN    The Creole's good here, if you...
MCU MAN
CAM  3       WOMAN  Sure.  OK.

             MAN    Yeah, if you think you'd...

SHOT 7       WOMAN  Yes. Fine; I'm sure, OK?
CU WOMAN
CAM  1

SHOT 8       MAN    Or what about the cajun chicken?
CU MAN
CAM  3

SHOT 9       WOMAN  I don't care.  OK?  Now how much
CU WOMAN            time do we have?
CAM  1
```

Figure 12.2 Shooting script

are multiple cameras in use, each one will be indicated individually, such as "CAM 1," "CAM 2," and so on. Each shot will be numbered and identified as "SHOT 1," for example). The shot types will be identified as follows:

ECU	extreme closeup
CU	closeup
MCU	medium closeup
MS	medium shot
MWS	medium wide shot
WS	wide shot
EWS	extreme wide shot
2S	two-shot (two heads in shot)
OS	over-shoulder shot
POV	point-of-view shot

Figure 12.3 Shot notation

Following shot type will be an identification of the people on things in the shot, and that will be followed by notation about any camera movement, such as

    SHOT 1
    CAM 1   EWS Talent A
                Slow zoom to:
                ECU Talent A

This system allows for brief and concise notation of virtually any shot composition and movement possibilities, and gives the designer a very specific mental image of the compositional parameters within which she must work.

It should be noted here that the shooting script, being the more closely defined format, gives the designer a more sharply delineated brief, and thus simplifies the process a bit. However, in the case of a drama script or when a more experienced designer is involved, this may be detrimental to the production. Often, the designer may have good suggestions about ways to achieve an effect in less time or with less expense and thus save the production a significant amount of money. The designer may have ideas or criticisms that the author and director didn't anticipate and may make suggestions that improve the quality of the piece. (Jo Mielziner, the legendary stage designer, was often called into the production process early on for this reason; the flashback scenes in *Death of a Salesman* were reputed to have been added after he explained how a scrim lighting effect could be used to make them work.)

As you can see, although it simplifies the production design process when the designer is given a shooting script to work with, it doesn't mean that this is necessarily the best course to follow.

Analysis

Often the toughest job for the designer, script analysis is also absolutely pivotal to the success of the process. I've noted that many directors are not visual people; this is because often they come up from the performance or management side of the business. Traditionally, directors will have been on-air talent or production assistants before moving up the ranks and will have little or no visual training and some innate fear of things visual and technical. It may not be entirely fair (what is?), but, in most cases, the full weight of the task of translating words to pictures will rest on the production designer's shoulders. This means that he must read the script (usually quickly) and absorb the visual imagery inherent in it. Some scriptwriters give detailed descriptions of the locales as they imagine them and others dwell on more esoteric things or give no visual clues at all. The early stages of script analysis, then, may vary from simply executing in a soft medium (such as pencil or marker or pastels) an impression of an already detailed description to imposing a very personal construction almost from the ground up.

At a more mundane level, every production design needs to establish certain fundamental factors about the program. The context of the presentation is the most essential element which is established visually: The nature of the piece and the relationships of its participants to each other and to the viewer can all be established and underscored by the visual treatment. This should be self-evident in the case of a scripted dramatic program; it is also possible and important for other formats, such as news, game, chat, or other entertainments. We dealt with this extensively in Chapter 10, Interpretation; What is relevant here is that all of the important decisions about program interpretation (where design is concerned) are made in the script-analysis phase of production. This may be a matter of days or it may stretch to months (if the budget permits).

When you have months to do it, this phase of planning can be quite enjoyable; when you have to turn it around overnight, it can be downright nerve wracking. I find that it often helps to scan the script (having read it thoroughly several times) looking for adjectives that may add up to a feeling or a mood from which to develop some visual ideas. Of course, there is nothing more helpful than a concrete context upon which to build; a news program set in a wild-West saloon would be ludicrous, but very easy to visualize. (Don't rule this out; I have done very serious corporate presentations in front of giant cacti and overscale longhorn skulls carved from plastic foam.)

Many times, the best way to start the creative process is to immerse yourself in the genre: Hours spent watching soup commercials might cause

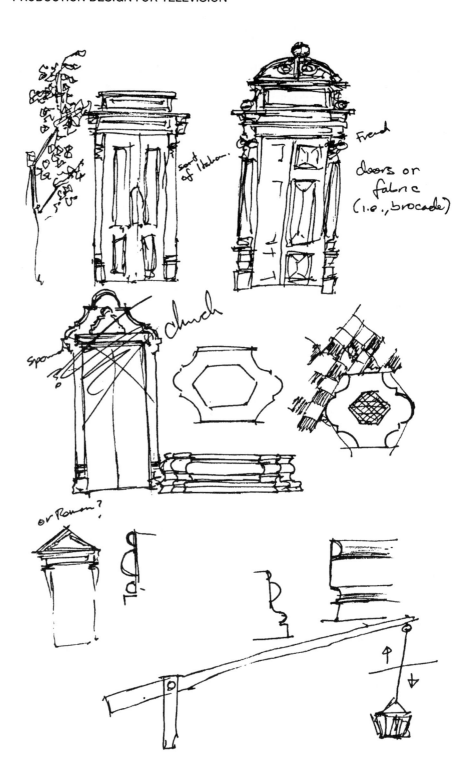

Figure 12.4 Some preliminary rough sketches

you to think about what you most dislike in reference to them, and thus coalesce your own ideas by defining what you *don'}t* want yours to be.

Whatever the mechanism by which you get into your script and make that leap from the written word to the first visualizations, this is a relationship with yourself that you must explore in order to shortcut the process when the project demands it. Few people in the television business really believe that what they are doing is high art; the budget and the deadline are sacrosanct, and few designers who show up at a production meeting empty-handed and claiming that they haven't had enough time to get inspired are hired again.

Storyboards

The point at which the words and visuals meet, prior to showing up on tape together, is the storyboard. Like the shooting script, this has a standard format which is universally employed. A page of storyboards consists of several rows of rectangles arranged in pairs. The top one of each pair is (coincidentally enough) 3×4, and represents the screen. Below each of these is another rectangle, usually more squat than the 3×4, which will contain the information relevant to each shot. This information may simply be a passage from the script or it may be a description of the type of shot and the manner in which it will be used.

The storyboard is a component part of the evolution of the design and as such will be revised several times at least before the design is finalized. Initially, the production designer will do working storyboards with monochrome line drawings in the frames. Later, when the look of the program has been finalized and agreed upon by all the parties who have to sign off on it, the designer will go ahead and do the finished storyboards.

Because of time constraints, the difference between working storyboards and finished ones may be little more than pencil versus ink. If the production is a large one, the sheer weight of work involved may preclude the sort of absolutely detailed storyboarding that is typically done for a less lengthy project such as a commercial or a newsbroadcast. In cases such as this, it is often sufficient to show the heads in shot, or the figures, and a few key elements of the background.

However, for commercials, in which the exact representation of detail can be regarded very critically, the designer may do presentation storyboards. These are as much a selling tool as a working aid. Typically, they will be rendered in color with markers or pencil, and are very accurate depictions of the intended final product. The client in these cases will regard the presentation storyboard as a contract document and will expect the final shots to look exactly the same.

In this case, finish is very important. At the highend of the market, presentation storyboards may be done in a photorealist style or in a glitzy airbrushed graphic manner (especially to illustrate fancy special effects).The storyboard is indicative of the difference between styles of production; for programs, the skills and talents of the artistic staff are relied on heavily to look after the details, whereas for commercials, the tendency is to spend much more time and money on preproduction planning and speci-

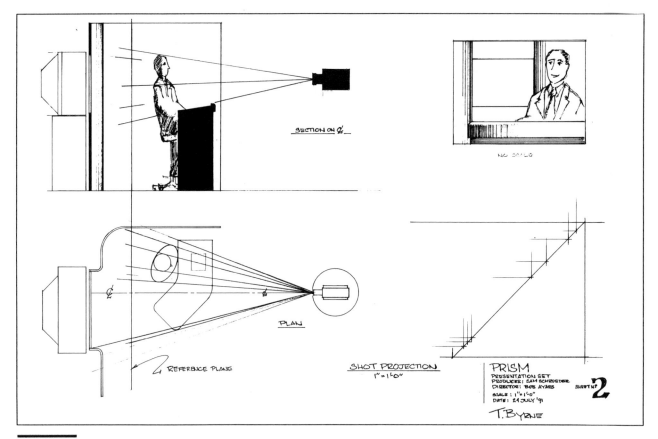

Figure 12.5 Floor plan and shot section used to plot and project the shot. Note how the places at which the plotted lines of important points in the shot (talent's shoulders, edges of in-shot monitor, etc.) cross the reference plane are carried across to the line on the right (45° angle) and up to the frame above.

fication of every detail. It is not unknown for a highly polished presentation to sell a commercial concept that is inherently inferior to another that may not have been presented to the same state of finish.

Projection of a shot is a complicated mental calculation, and there are means and devices at hand to help you with it. Remember that the television camera is intended to move about; it's usually mounted on a pedestal that can roll around the studio floor and moves easily up and down. A reasonably agile camera operator can, probably will, achieve an amazing variety of shot angles. As the designer, you will need to know where the camera is placed in relation to the people and things in a shot (both in the horizontal and vertical planes) whether it is static or moving and, if mobile, which direction it's going and where its travel begins and ends.

Initially, shot content is best calculated by drawing a floor plan and a shot section. The *foor plan* (sometimes called a *ground plan*, though this is more an architectural term than a television one) is simply an accurately

---

The design should evolve from its initial expression in terms of what the camera will see. The storyboard is simply a commitment to what the camera will see. This is another aspect of production design that only experience can provide. After many hours of looking through the camera viewfinders and control-room monitors, you will develop the natural ability to size up everything you see in terms of the shot.

scaled aerial view of your studio setup. This will give you the real-size relative proportions and positions of your set pieces, talent, and cameras and thus will allow you to determine realistically where the cameras can be placed and what shots are likely to be used.

The *shot section* is a vertical slice through each shot, indicating the height of the camera lens relative to the talent and set, and (in combination with the plan) will give you a clear and accurate idea of everything that will be in shot.

There is a device called a *shot plotter*, which can assist you. The shot plotter is like a protractor, which will allow you to plot the exact angle of a particular shot and thus accurately determine what will be seen in the shot. Since the advent of zoom lenses, these plotters have become somewhat more scarce because the only parameters of a zoom are the extreme wide-shot limits. In the old days of turret lenses, the shot angles were fixed and the use of a shot plotter was critical. Still, it's a handy device, and allows the designer to calculate the 3:4 aspect ratio, which rules every shot, quickly and easily. With these tools available, it is possible for the novice designer to calculate each and every shot, and render it accurately in the storyboard. The more experienced designer will usually not need to do this, but that's not to say that he never will. Often, the specific dimensions of a shot are critical to its function, and such exact measurement and calculation can't be fudged. In a shot with a number of piled-on elements, such as the typical news anchor shot, the exact division of the frame is critical and requires a lot of math and geometry from the designer.

This may be true of other types of shots, such as commercials, but we'll use the news setup as an example. The electronically constructed graphic inserts are the pivotal elements of these shots because, although they are adjustable in size and shape, the dimensions must be locked in at some point (before the show goes on air) and kept there for the duration. These elements would include the over-shoulder boxes we've come to know and love and the identifying information so often inserted across the bottom of the frame. Each of these items eats up a certain amount of precious picture area and (once locked in) can't be adjusted to allow more room for the talking head.

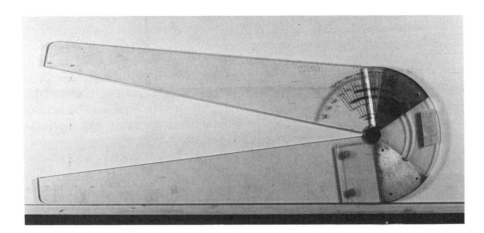

Figure 12.6 A shot plotter

In such a situation, it is the designer's job to plot the exact dimensions of each bit of the picture, so that the people who must assemble the puzzle can do it accurately and consistently night after night. This is no small task, and will demand skillful use of geometry, draftsmanship, and tact.

To begin with, you will want to decide how much of the frame should be consumed by the graphic box. This is where the tact comes in; often, the director and the graphic designer will have in mind a rather large box literally chockfull of valuable information. This is because many news directors are really print journalists at heart and because all graphic designers feel themselves to be neglected. As the production designer, it will fall to you to defend the overall shot composition and the need to clearly see the face of the talent. These shots work best when the talent is the dominant image and doesn't look like someone peering around a corner, and you will likely have to defend this logic. Having sorted that out, you'll need to decide how many other graphic elements will be inserted at any point or under any contingent circumstances. It is never sufficient just to design for the usual configuration; the creative team should do its best to anticipate every possible contingency and allow for it. Consistency and continuity are the marks of good design and good program planning, and will equal credibility in the minds of the audience.

Having made these decisions, you can now move on to the shot elements that are somewhat more flexible. The talking head is, of necessity, one of these. People, unfortunately, come in a wide variety of shapes and sizes. Some are thin, some are fat, some have no hair, and others have massive constructions surrounding their faces. As a rule, it's best to discourage large hairdos, because they tend to force some of the head to fall behind the graphic box, or cause the head to be greatly reduced in shot, making that particular person look like a midget next to the others. This is an aspect of vanity where the talent is concerned, and you may expect a pitched battle when you ask one to change the hairdo. The optimum placement of the head is with the eyeline about one third of the way down from the top of the frame, and the bottom of the shot above the elbow line, so as not to see the hands fidgeting with a pencil or shuffling script pages on the desktop. Sometimes, the only way to achieve this is to bury the hair behind the graphics, but it's a poor solution and a major compromise.(Fortunately, as I write this, the "big hair" look is fading out of fashion!)

Having ascertained that the difficult shot with the over-shoulder box is going to work, we can move on to the shot that inevitably follows it in practice—the single head centered in the frame. To achieve this, we usually fade the graphic insert out and pan the camera left or right until the head is in the center of the frame. Having imagined this move, the designer now has to visualize the shot without graphic inserts: how big the head is in the frame, how high the camera is in relation to the talent and the desktop, and how much (precisely!) of the background is in the shot. At this point will come consideration of the details to be designed into the background treatment. Will there be an identifying logo? How much of the background will be in focus? Will the color, texture, and identifying marks on the background be consistent with and supportive of the graphic treatment that immediately preceded, or will the transition be too jarring?

A treatment that is often used in cases such as this is to keep the shot composition asymmetrical after the graphic box is removed and place a design device of some kind in the space to the side of the frame where the box had been and so retain the compositional balance. This could be a program logo, an in-shot monitor, or a set element such as a plant or a picture hung on the wall.

Now the easy part is done and our designer can move onto the messy part. At some point in any news program, the director is going to want multiple-head shots to do interviews or to show several presenters chatting about the weather forecast or the latest gruesome airplane crash. There are many possibilities; in order to plan for them, the designer must take the time to discuss the subject extensively with the director and try to anticipate all of the possible combinations the show might require.

Again, we will begin with the shots and then try various placements of bodies which will facilitate them. If we've decided to have two news anchors, a weather reporter, and a sports reporter, we need to find places to put them that will allow for the single head shot with graphics, a clean head shot, a group shot (four-shot), and two-shots in any combination to allow a hand-off from any one to the adjacent head on either side. The difficulty with these objectives is to come up with shots that will compose well in the 3x4 format without leaving too much dead space above and below the heads and without making them seem as though they are actually miles away from each other. Often, the best way to achieve this is to put the camera off-center and shoot the four-shot from a fairly extreme side angle. This makes for a more interesting composition and gives an added appearance of depth. (Remember that one of the objectives of television design is to give the appearance of three dimensions in a two-dimensional medium.)

Let's say we've achieved this through careful drafting of floor plans and accurate shot plotting. At this point, the director usually says, "Oh, and I'd like to get some over-shoulder shots for the interview segments." To meet this criterion, we design a separate interview area. A good solution to all of these requirements is that achieved for the WNBC news in New York, incorporating a four-handed news desk and an interview area that is largely open, allowing the cameras to circle around the guest and interviewer, and using the news desk (and its occupants) as background dressing. (This also facilitates the hand-off going into the interview and leaving it to return to the anchor positions.) Such interview positions are often described in a kind of mathematical-formula shorthand as a *one-plus-one* or *one-plus-two*, for example, the "one" being the interviewer, and the other number the guest or guests.

The nature of a given production will dictate the detail with which the shots and other visual elements are planned and the meticulousness with which the plans are followed. A dramatic program will tend to allow more scope for creative adjustment during the shoot and thus require a somewhat less meticulous storyboarding process. A highly graphic piece, such as a

---

A good television designer never designs for the wide shot, which typically falls last in the planning process. The wide shot is often used only at the end of the program; has credits rolling over it; and occupies perhaps 30 seconds of air time.

Figure 12.7 The one-plus-one interview area from WNBC-TV News. Design: Frank Schneider

commercial or a news show, which typically will involve the assembly of a number of elements from different sources, will require tighter controls on shot planning and tighter adherence to the storyboards in the execution.

The camera is the critical element of the medium; it decides what we see and influences our interpretation of it. The good production designer will know what the camera can do and will learn to control it; this is the most effective tool in the design arsenal, and the storyboard is the best way to understand what it will do for us and exercise control over that. The designer is the link between the verbal expression of an idea and its visual realization; the ability to understand and interpret ideas from the script and define them visually through the storyboard is pivotal to the whole process of television production.

Rendering and Developing the Idea

13 ▫▫▫▫

Rendering as a Process

At some point in my career, I stopped to consider the meaning of the term *rendering*. I'd always used it as a noun: A rendering was the finished presentation of the design idea, usually done in gouache paints on illustration board and nicely matted.

Television is a collaborative art and, as such, is a messy one. The development of the designs required for a given project is a process of evolution—one of presentation, rejection, modification, and, finally, solution. The solution is never definitive; there is no such thing. A given production will achieve the best solution possible for that particular group under those particular circumstances at that particular point in time. This is a good thing; part of the fun of doing this sort of work is the reinterpretation of an idea: putting one's own spin on an old idea. Whether it's Shakespeare or "Wheel of Fortune," a new creative group assigned to the project will probably want to put its own mark on the work, and it is natural that they should.

Notice that I refer often to the creative group or to creatives in the plural. We mustn't forget that this is a collaborative process, and that the designer is never working alone (or ought not to be). The development of the program idea is a process and the development of the design idea (as a component of that larger work) must be collaborative as well. *Rendering*, in this context, is now a verb rather than a noun; rendering is a function of the development process, a means of communicating the visual aspect of the production idea as it progresses through the various stages of development.

Preliminary Rendering Techniques

As noted, the sequence of events that neatly lays out the progression of the design idea is more than a bit naive. It is not realistic to attempt to force any such process into a sequence, except to say that you begin with a vague idea and end up with a specific one.

Where preliminary rendering is concerned, however, certain rules of expediency tend to apply. The time frame for development of ideas (for the creative part of the process) is never generous and rarely adequate. Also, the ideas at the early stages tend to be vague and tend to demand a soft me-

Figure 13.1 A soft sketch

dium. For these reasons, many designers will work in media such as pencil, charcoal, or marker, all of which are fast and soft and permit rapid revision. Also, these media represent less of a commitment to the specific placement of line and form and are more prone to happy accidents such as an inadvertent smudge of a line or bleed-through of colors (which have proved themselves a source of inspiration in more cases than most designers choose to admit).

You may note that I assume the ability to draw. Many people who haven't mastered the skills of draftsmanship call themselves designers—I don't. Anybody can have wonderful ideas; this is no special gift, and many people regularly fall asleep at night and see the most complex and fascinating visual constructions in their minds. A designer, however, is someone who can manipulate those ideas toward a specific objective and can render them physically and accurately at every stage of the creative process, so that others can see them. Anyone without the hand/brain coordination to put ideas on paper or on a visual display unit (VDU) screen is just another person with a fanciful imagination, and not a designer.

I'll proceed on the assumption that those who have gotten this far will have the required drawing skills and understand what I mean when I talk about "hard" and "soft" media and commitment to line and form. Usually, the progression we're talking about here is from soft to hard, from vague to specific, noncommittal to definitive. So, while one may begin with charcoal or soft pencil, as the ideas take on more specific form, the medium may change to a harder one such as pen and ink or even paint. (I will continue to refer to these media by their traditional names because most computer-controlled graphic paint systems are designed to mimic the conventional media. If you're using such an electronic system, you will no doubt understand me.) The trend will also be to move from monochrome to color, incorporating marker washes, watercolor, pastels, or colored pencils.

How you choose to show the design will vary according to personal preference and the demands of the specific project and the other creative personnel. Much of the visual information, as noted, will be shown in storyboard format, shot by shot. Chances are that you will have neither the time nor the inclination to render each storyboard frame in color and great detail; few projects allow enough time or pay well enough to justify this kind of effort.

Typically, the storyboards will be done in monochrome, often hard line, and often by a designer's assistant. The full treatment will be reserved for a single wide-shot view of the studio and used to sell the producer on the idea rather than as a tool of accurate representation from which to build and finish the setting.

Evolution of the Idea

The bulk of the effort, and time, in the rendering department will be spent on a single finished color rendition or a model of the studio.

The process by which we arrive at this point may, depending on the specifics of the project, have required the development of a comprehensive

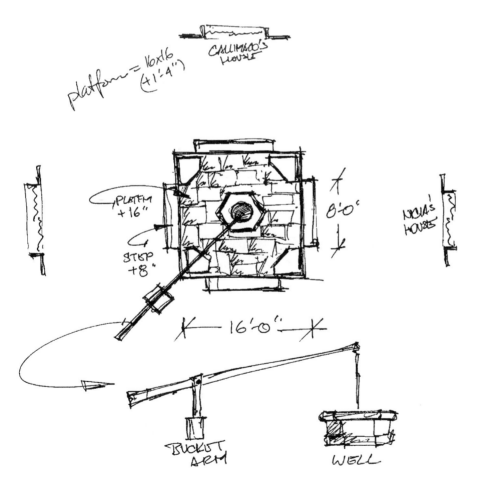

Figure 13.2 A rough-sketch floor plan

set of rough storyboards, a set of plans and elevations, and perhaps one or more models. If it is the news program described earlier, it will have required the most exact possible roughs in order to enable us to plot the shots with some precision and thus to calculate exact placement of cameras and shot composition. We have to know for certain that the elements of each shot are going to fall exactly where we want them, so we can alter them if they don't. This may seem like a lot of work up front, but the rule that applies in this case is that it's always easier to change the set *before* it's built!

On a show such as this, the process often requires that the designer begin with the floor plan and develop the look of the set from there, having calculated the camera positions and studio traffic patterns before getting to the creative bits. It never pays to ignore the mundane aspects of studio design.

Many production designers (myself included) find that the most useful and, oddly, the most flexible plan is to develop a preliminary floor plan and build a model on top of that. This may seem like a lot of work for an early stage of the design, but the same axiom applies: It's easier to change the model than to change the set.

Model-Making

This is a good place to define some terms and practices with regard to model making. Most models are made by the designer or an assistant, and the skilled model-maker may find this a separate career, or at least a good entry-level position with which to begin a career leading to production design.

Most models are made of cardboard. I use cold-press illustration board, which is available in single and double thicknesses. The double weight is good for the base of the model and for any walls of the studio or parts which may need to be a bit stronger. Cold-press board takes paint very well and gives a matte finish; the hot-press surface is designed for ink techniques, and is a bit too shiny for my taste. Foam-core board is also useful: It's lightweight and very strong, but can tend to curl and, once curled, will not straighten out again.

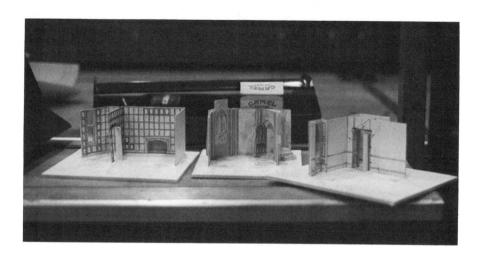

Figure 13.3 A white model. Shot on the studio floor at BBC Wood Lane Studios, London, during a load-in (fit-up). Designer unknown

Plate 1 A set being erected in studio. Radio Telefis Eireann, Dublin. Design: Alan Farquharson

Plate 2 The same set dressed and lit. Lighting: Bob Moore

Plate 3 Lighting used to add color to a neutral set. Lighting: Bob Moore, RTE, Dublin

Plate 4 Period costumes. Since they often include more extravagant makeup and wigs, period costumes can be more fun and more challenging to the costume designer. Design: Sue Wilson

Plate 5 WNBC news set showing blue CSO panel. Design: Frank Schneider

Plate 6 The art of the scene painter: A painted cyclorama. BBC Wood Lane Studios, London

Plate 7 The art of the scene painter: Painted vista with low-relief foreground pieces

Plate 8 The finished effect seen through the set window

Plate 9 A finished rendering: Shadow of a gunman

Plate 10 An over-shoulder news graphic. KYW-TV, Design: Adrianne Kerchner

Plate 11 The KYW-TV news set showing green CSO panels

Plate 12 KYW-TV news weather position in use

Plate 13 A set on the studio floor. RTE, Dublin

Plate 14 The same set as seen in shot. The shot is all that matters!

Plate 15 Focus: Using color to draw attention

Plate 16 A set on the scene shop floor. Note the use of real wallpaper and the way that out-of-shot stairways are constructed. RTE, Dublin

Plate 17 Setting up a tracking shot on location. A simple camera move can be very labor intensive out of studio. Glenroe: RTE, On location in County Wicklow

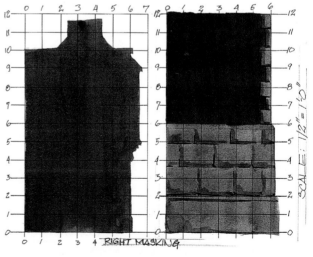

Plate 18 Paint elevations

Plate 19 Paint elevations. Note the scale grid marked over the painted color elevation to assist the scenic artists in laying out.

One early-stage type of model (for which foam-core board is ideally suited) is the white model. This is typically a soft presentation: the measurements may not be quite as accurate as on a finished model and it is usually done without any color or details of finish. Even so, it can be extremely helpful in setting up possible shots and placing set elements in the studio. Often, the set pieces may be made in sections and not affixed to the base of the model, to allow the director and the designer to move them around and play with different arrangements until they come up with a workable solution. They may also construct a viewing box from cardboard, representing the viewing angle of a typical shot (calculated from the shot plotting that was done on the preliminary floor plan).

In any case, the models will be made to scale. This means that a consistent scale is adopted and adhered to throughout. This may be ½" to the foot (in which case ½" on the model corresponds to a foot in reality) or ¼", or even ⅛". Unless the studio is immense, the usual scale is ½", because this allows the director to get her eye right down into the model and more accurately visualize each shot. The model, at this stage, is regarded in the same way as a rough sketch. This may be quite painful in the early stages, but a designer should be prepared to throw away a model without much more anguish than she would tear a page from a sketchbook. Remember that the product of your labors is the finished set in the studio and all of your drawings and models are merely steps in the process by which that final result will come to fruition. If you can't bear to throw it out, give the thing to a neighbor's kid to play with. I no longer keep even finished models; I set them up, light them, and take several nice photos, then throw them out—if I'd kept them I would have room for nothing else!

Rendering on Location

If you happen to be working on location, a different set of rules and a different process apply. Since you are not working with a blank slate and you're able to exercise much less control over the situation than in the studio, the rendering process will begin with the documentation of the existing locales. This is usually done photographically, on a visit to the locale (known as a *scout*. (The British term is *recce* , presumably short for *reconnoitre*.)

On the scout, the designer will carry a still camera and lots of color film. Standing in a central position (preferably where the camera is likely to be), he will shoot a series of panoramic shots. By holding the camera very steady and slowly rotating between exposures, it is possible to come away with a set of pictures that can be glued side-by-side and slightly overlapping onto a piece of cardboard to make a sweeping panoramic vista. It may be a bit rough here and there, but it will serve as a wonderfully useful reference about the location's features.

This location photomontage may be accompanied by a rough but tolerably accurate plan of the site, which will show the actual relationships of the things in shot and their relative sizes. This site plan will permit the designer and the director to plot their shots as they would do in studio. The designer then does storyboards indicating the compositional content of

Figure 13.4 Location photomontage. Sequential photos stuck together to make a panorama

each shot and thus decides what dressing elements will be required for the shoot. It should be noted that location shoots are, by necessity, somewhat sloppier than studio shoots, and the shot plans made for them cannot be relied on to the same degree of accuracy as those for studio shots. For this reason, the designer must allow a bit more slop in the dressing plans, assuming that the shots may include more of the location than originally planned, and must develop contingency plans to cover this.

On location, most of the design is simply dressing existing buildings. The reason for choosing a location is that it already looks somewhat like the desired setting for the scene, and using the existing site will save money and time. Thus, the rendering process for location design is usually limited to drawings and photographs. A location may need some special signs, different curtains in a window, or perhaps a particular car in shot. In cases such as these, a set of scout photos and a sketch of the portion of the location scene to be used in the shoot, showing the items of set dressing in place, will usually prove sufficient. (Rarely will a director do a location shoot without having visited the location first, and anyone who cannot visualize the scene when it is presented full-size really ought not to be directing.)

It is rare for a model to be made of a location, but not inconceivable that it may be useful to do so in special circumstances. Of course, when exterior sets are being built on an outdoor back lot rather than in a studio, the process is virtually the same as if the set were being erected in a studio. The basic difference between studio and location production is one of control; the studio is a friendly environment in which everyone is working toward the same goal, while the location is hostile in that there are more distractions and potential hazards such as weather and traffic. For these reasons, locations are also places in which time is of the essence; since it often requires the stopping or redirection of traffic and interruption of commerce, a location shoot must happen quickly and efficiently. Thus it is necessary to plan location work thoroughly and well in advance, and the preproduction visualization process becomes even more important.

The Final Rendering

I've stressed the importance of groundwork; the need to do accurate floor plans and to plot shots first, before addressing the more creative parts of the process, has been restated again and again. One advantage of this method of working is credibility; if you do your homework and plan your shots accurately, the final product will actually resemble your drawings quite closely and everyone will be impressed. (In addition, you may be hired for that company's next project!)

The final rendering phase, however, is when you get to show your creative stuff. The object of the final rendered drawing or rendered model is to capture the look of the show in all its excitement and dynamism. This is easier to do in a sketch than in a model and sometimes you will decide to do both. A model cannot show accurately the way the set will look under studio lighting. If your painting and drawing skills are sufficiently advanced, you will have no trouble rendering the atmosphere of the light and the effect of camera-lens distortion (or even embellishing them a bit).

Apart from being a tool to finally sell your idea to the people who have to pay for it, the rendering or model will also likely be the thing you include in your portfolio when you go out pounding the pavement to find your next job.

As I mentioned, I tend to use pencil for a soft medium and progress to watercolor and gouache for the final rendering. There are several reasons. The first is that I learned to render this way and feel most comfortable working in these media. Another person may have learned on markers, and thus may find that a more natural tool. I like the translucent quality of water color, and the capacity to overpaint with gouache in those areas where

Figure 13.5 A finished rendering

Figure 13.6 A rendered model

I want a more solid feel. Also, it is another means of progressing from soft to hard media: I may begin with a light pencil outline and add in washes of translucent color (often dyes, which give a really luminous effect) for the background areas, which would typically represent lighting effects on a *cyclorama*. Foreground objects can then be drawn in with a heavier pencil line, and painted in with gouache colors, to give them a solid feel. Finally, the people are included, again in gouache, then atmosphere effects such as beams of light and subtler shadings of various forms can be laid in with an airbrush. If necessary, I sometimes follow this with a few lines of ink for definition here and there.

The object here is to point up features such as color, form, and lighting to sell the idea and to communicate with and motivate the other creative personnel, such as the camera crew and the lighting director. The final rendering should be as polished as you are capable of making it and should be a real presentation piece: A clean matte, a neat cover sheet, and a piece of clear acetate will impress those who need to be so impressed, and will enhance your credibility with everyone else on the team.

If you're doing a rendered model, you will need to have a good memory and an eye for textures and objects from everyday life that can be adapted to your models. As I noted before, you will probably find cold-press illustration board a good material for model making. While this is good for basic work, and takes paint well, there are going to be surface textures that you'll want to portray in ways other than by trying to imitate them with paint techniques.

If, for instance, you designed a setting with carpets on some of the platforms and lighting trusses overhead, you would need to improvise. The carpets can be represented by finding a fabric of suitable texture (keeping in mind the scale of the model) in exactly the right color to match your carpet samples. If you can't get the color right, take some white fabric, and paint it. To do lighting trusses, I have used a very realistic (if time-consuming) technique. I get brazing rods (they're available *very* thin) and solder them into a scale box truss configuration. Using brass tube bought at a hobby shop, I make little lighting instruments, which I solder into the truss framework at various angles approximating the positions of real studio lighting. The whole thing can then be sprayed black or silver and glued onto the model.

Furniture can be made by building the basic shape from cardboard and sculpting over that with acrylic modeling paste. Scale people can also be made using this technique, which takes paint fairly well and dries fairly quickly. Weeds can be used to represent trees, electrical wire and plastic beads can make chandeliers, and draped fabric can be represented by Kleenex painted with diluted glue. I often add acrylic matte medium to gouache paint to prevent bleeding through and to protect the model from damage by abrasion and moisture.

Model-making requires a fertile imagination and a singleness of purpose: I have made impressions of women's jewelry in clay, then molded a copy in soft rubber to make a pattern stamp for wallpaper textures. Your friends may think you've become very strange, but once you get into this, it's quite enjoyable and a bit of a challenge. It also requires a steady hand and a set of fine brushes for the detailed parts.

Access to a reducing copier is good, also; many settings require in-shot graphic treatments and the quickest way to generate them is through a succession of reductions. Once you've got the graphic or logo (or whatever) down to the right size, you can trace it onto the model and paint it in. There are other, more accurate methods, and if you have access to them, use them. However, I've found that, as a freelance designer, I rarely had the money to cover the cost (these services tend to be expensive) and often found myself working on a model late at night and unwilling to wait until the next day to finish up.

As the representation of creative input to the project, your rendering will be the single piece of work that most strongly documents your personal design style. It will also, in the long term, represent your talent in your quest for further employment. In either context, you want it to show your creative ability to best advantage and, as such, you should see to it that the rendering of your ideas is done to the highest level of finish and detail of which you are capable. The ability to render your ideas with accuracy and energy is one of the essential components of the designer's craft and must be mastered.

14 Plots and Plans

Documentation Drawings

Plots and *plans* refer not to secret machinations but to documentation: Plotting and planning are the processes of preparing execution drawings for the construction and installation of scenery and lighting in the studio or on location.

These drawings are an essential part of the process; the exact details of the design are expressed in them and they form a part of the contract between the designer and the subcontractors who supply the finished sets, props, and sometimes lighting. Accuracy and thoroughness are the watchwords in the preparation of execution drawings, because any errors in them can be costly if not discovered before the piece is built or the lighting equipment hung, and the responsibility for this falls to the designer. There are certain categories within which these drawings may be classified and certain sets of rules by which their format may be defined.

Floor Plans and Site Plans

As I noted previously, the floor plan is often the first drawing generated (even if it's only a rough one) and nearly always the first measured drawing. As with all execution drawings, these are done to a specific scale and are so noted.

The floor plan is simply an aerial view of the studio with the set shown in its proper position. You may like to think of the floor plan as a map; that's really what it is. As a map, it often incorporates a measured grid—a set of lines, perpendicular to each other, approximately 3 feet apart, from which any location on the plan can be found. Often, the studio floor itself will have this grid marked out on it (typically etched into the floor so as not to show up in shot) to make placement of set pieces quick, easy, and accurate.

A floor plan will typically be done to the scale of ¼"=1' or sometimes smaller, if the studio is very large. If it is carefully measured by the draftsperson and a grid is marked out on it, there is not usually any need to indicate specific dimensions. There is one exception, and this is in certain instances of height: The height of a platform (and the steps leading up to it) should be shown and the height of an opening (such as a doorway), if there

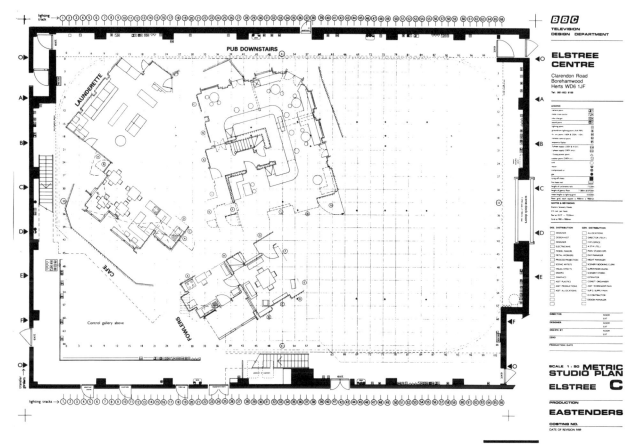

Figure 14.1 A studio floor plan: BBC's Eastenders. Design: Nigel Curzon

is any reason to believe that such information may be useful, should be given.

The people using this drawing (apart from the designer, of course) will be the director, the staging crew, the carpentry crew, and the studio crew heads, such as the lighting, camera, and sound people. If any information is thought to be useful or necessary to them in the completion of their tasks, it should be noted on this drawing, unless (as in the case of the construction crew) they are to receive other drawings that include information not necessary to the other personnel.

The site plans drawn up for location shoots are conceptually the same as for the studio floor plan. On location, it may not be feasible to do accurate measurements of every building in the area, although every attempt should be made to get exact dimensions of those parts actually used in the process of dressing the site. If a sign is to be hung on a building, we need to know whether it covers an existing one, how it is to be supported, and exactly how big to make it. This would require accurate measurement of the lateral and vertical dimensions of the building wall, and the placement and size of windows, doors, and so on.

Elevations and Details

This brings us into the other dimension: height. In cases in which details of vertical surfaces are important (most cases) we revert to a drawing known as an *elevation* to show these dimensions and details.

This may show a piece to be built for studio use or it may show the vertical surface of something to be dressed on location; in either case, it is called an elevation and contains the same type of information. An elevation is not a *perspective drawing*: In an elevation, we merely show a flattened-out picture of each vertical surface; in a perspective drawing, the appearance of lines converging in the distance would convey a sense of three dimensions. The elevation, like the plan, is more of a map or chart than a representation of the way a thing might look in three dimensions. There is a good reason for this: The elevation is a *dimensioned* drawing, in which every critical measurement (and some that may not be critical) will be included according to the accepted dimensioning style.

These dimensions will typically be overall outside dimensions, height and width of doors and windows, placement of moldings on walls, height of platforms, locations of logos or other artwork, and other information of that nature. These drawings may also contain (within them) other drawings, done to a larger scale, of certain critical items of construction or finish. These drawings are called *details* and are usually depictions of smaller measurements or specifics of construction or finish. Examples would be

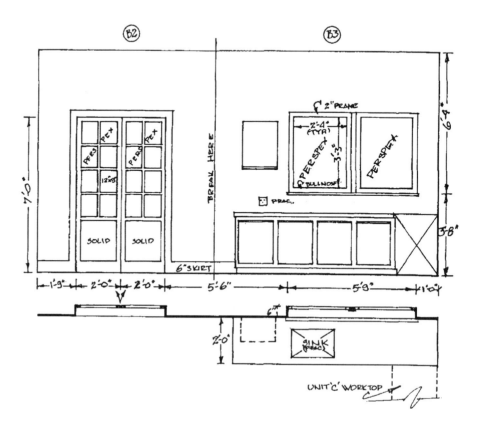

Figure 14.2 Elevations: Glenroe (RTE)

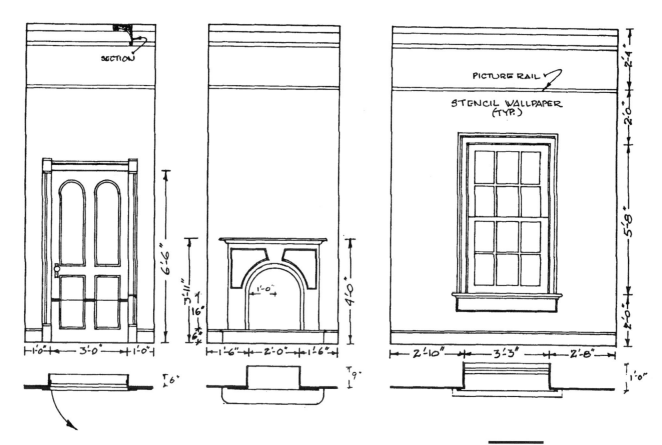

Figure 14.3 Details: A door, a fireplace, and a window

door or window construction details, mounting of hardware such as doorknobs or of joining hardware.

Another sort of elevation is useful for showing details of internal construction that may not be visible to the eye when the piece is completed. This is called the *section*, and is literally a slice through the piece of scenery as though we had attacked it with a chainsaw. There is a specific accepted method of identifying and labeling sections, which indicates precisely where the imaginary cut is made and which portion of the piece is left standing for us to see. This is only a *theoretical* slice through the set; make sure no one actually cuts it!

The elevations will also be filled with marginal notations, explaining to the carpenters just what materials are specified, which moldings to use, and what surface finishes are desired. It is a good practice to include a plan view of each piece to show such things as which way doors open and how the different pieces fit together when they are assembled in the studio.

Additionally, when there is time, you will do color (or painters') elevations. These will omit details of carpentry and dimensions unrelated to the final surface finishes. Dimensions relating to the placement of painted designs on the surface of the built piece will be included. The whole drawing should be covered with a clear piece of acetate, on which is inked a grid of lines 1 foot apart (in scale). This will permit the painters to chalk in ref-

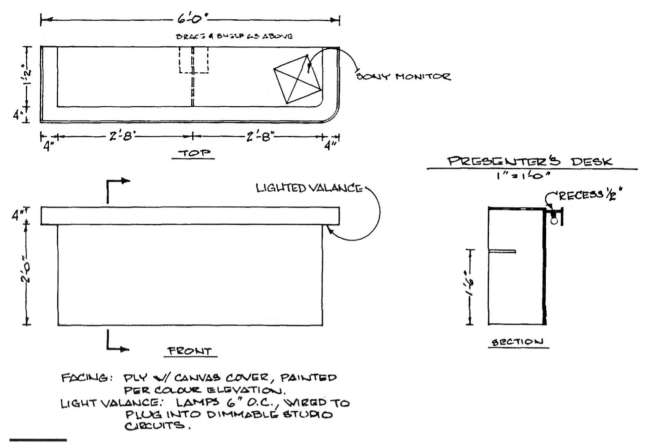

Figure 14.4 Details: A desk for a game show

erence lines and use them to place details of finish on the full-size piece, and the acetate will help to keep drips and slops of paint from ruining your slaved-over paint elevations.

If surface finishes are not painted, it is sometimes sufficient to specify brand name, manufacturer's catalogue number, or other identifying descriptions. If it is possible, you may want to attach sample pieces or cuttings (if, for instance, the covering is fabric or wallpaper) together with identifying information. It is a common practice, when time permits, to render the colors into the elevation even when the surface is fabric or formica for your own satisfaction—to confirm one last time that the color choices were the right ones and it really is going to look the way you want it to.

Plots

We'll begin with the lighting plot (or plan) since that is the most complicated and usually the most critical to the work of the production designer. In essence, the lighting plot may be a simple, unscaled sketch: Many a film lighting director (gaffer) has submitted a light plot on the back of an envelope from a meeting with the director over coffee. On the whatever-works theory, this is alright. From a design point of view, it is less than adequate.

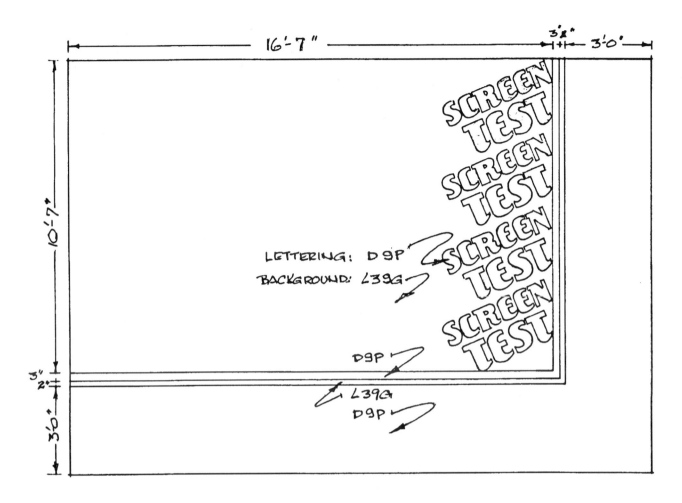

Figure 14.5 A paint elevation. Note that colors are indicated by stock numbers

The typical television studio setup is a bit more complicated than many film shoots and there are fundamental differences that make it so. Lighting in the film tradition is done from the floor, with equipment on rolling stands. The theory is to light for each shot and to light piecemeal in the same way that the scenes themselves are rehearsed and shot, using only what is necessary for each take.

Television, however, tends to go for the multiple-camera setup, which requires an overhead lighting grid and a large number of fixed lighting instruments, so that it is possible to go through a whole sequence of scenes in rapid succession, without stopping at each juncture to relight. Even if the piece is being shot *film style* (with only one camera), the normal practice in studio situations is to hang a plot from the grid.

At roughly the same time, they both may begin the installation process; while the set is being assembled on the floor, the lighting crew may be up on the grid positioning lighting equipment and making connections. (If the grid is accessible only from the floor, they may go in ahead of the set crew or after them.) When the set is assembled, the painters will move in to do touch-up work, and the lighting crew will begin focusing (aiming) the lighting equipment. In a large setup (fit-up to the British) for a complex multiple-set show, there may be as many lighting instruments as are used in a Broadway or West-End musical (as many as 300 or more). It is at this level that the light plot is indispensable; without it, the setup would be slow, riddled with errors, and very wasteful of expensive studio time.

With this in mind, the lighting designer proceeds to build her light plot. This typically begins with the production designer's floor plan, showing the studio and the various pieces of scenery in their playing positions. The lighting requirements will have been worked out in preproduction meetings, and the lighting designer will have already roughed in the plot or will have it in her head. At this point, she measures the surface areas to be lit, determines the type and number of instruments required to achieve the desired effects, and plans the placement of each source for the best angle. Vertical angles are also important and it may be necessary for the lighting designer to request a section drawing (or to rough one out herself) in order to calculate the correct placement of the lighting instruments.

Figure 14.6 A light plot

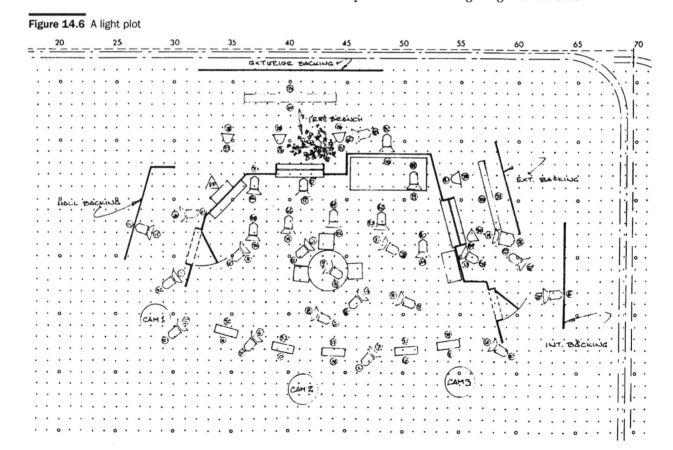

Having satisfactorily solved all of these problems, she will move on to specify color filters (gels) where they are required and to calculate connections to control circuits and electrical loads (so as not to blow circuit breakers or start any fires). The completed plot, then, will show exactly where to place each instrument, what it does, what color of light comes out of it, and where to connect it. In some instances, it may be left to the electricians to decide connections; however, it is always the designer's responsibility to know what quantity of equipment a given system can handle and not to exceed that.

To execute this drawing, there are many lighting templates available in every market, showing accurately scaled lighting equipment in a plan view (as seen from above). This simplifies the process of drawing and assures the designer that the instruments she specifies will actually fit in the space allocated.

Not unrelated to the light plot is the shot plot. As I noted, this may be developed in rough form at an early stage of planning, to assure the director (and the designer) that the set will provide adequate space for cameras and will yield the exact shots shown on the storyboards.

Just as the light plot illustrates the area filled by the cone of a beam of light, the camera shot plot indicates the cone represented by the *angle of acceptance* of a given shot. In other words, two lines are drawn, emanating from the front of the camera lens, indicating exactly how much of all that is in front of that camera will be included in the shot. (For optical purposes, this really is a cone; when it reaches the camera, it is masked off to make the old familiar 3 × 4 rectangle.)

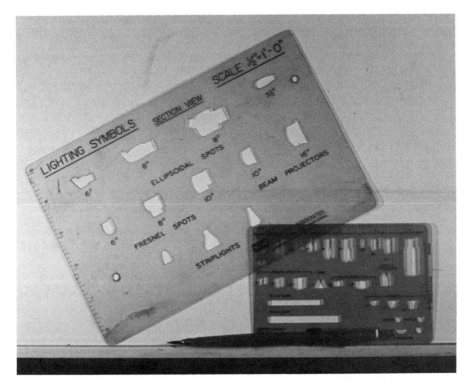

Figure 14.7 Some lighting templates. One is ¼" scale, the other ½".

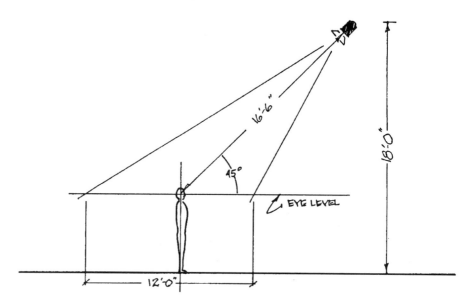

Figure 14.8 A lighting section. This shows how large the useful light pool will be.

In many instances, the rough shot plotting done to create the storyboards will be more than adequate for the whole process. In some cases, such as a news show, it may be necessary to plot the exact positions of the cameras on the studio floor. I recall tales of a director from the BBC's old-school training who would put chalk marks on the floor for the cameras and, from the control-room monitor, could tell if a camera was 6 inches off its mark! Many news studios are now installing robot cameras, controlled remotely from the production console. Once oriented to the studio, these can find exact positions anywhere on the floor. So, there are circumstances today in which the ability to plot your shots exactly and thus place cameras exactly is a useful skill, and the likelihood is that such instances may increase as higher technology invades the studio floor.

I've described the device called a shot plotter and explained that it was more useful before the zoom lens became standard. It is still handy to have, and speeds the process of drawing the shot plot just as the lighting template speeds the drawing of that document. Also similarly to the lighting plot, the shot plot often requires both plan and section views to give completely accurate information. It is a good idea to check height considerations as well as the placement of cameras in plan, both to make sure that what you want in view is, in fact, there and to ascertain that things which you do not want in the way will not be.

Drafting Technique

Drafting for theatre and television production design is a hybrid art. While technical drawing or engineering drawing is similar, it is much more tightly controlled and has a more closely defined set of symbols and rules than for theatre and television. Drafting for scenic art is somewhere between technical drawing and sketching, requiring both a fairly high degree of accuracy and an expression of individual artistic flair. This is asking a lot, but the best production design draftsmanship regularly achieves both objectives.

Media

There are two traditional media used in such TV/theatre drawing: pencil and ink. In truth, pencil is indispensable; those who draft in ink are really just laying in the final lines in ink, having done a number of pencil drawings beforehand. The difficulty of removing ink lines from paper without destroying the surface makes this a very poor origination medium.

For those with no experience of drafting, the pencil typically used is a mechanical (or knock) pencil; this is a lead holder with a spring-loaded jaw that clamps the lead in place. It is sharpened by spinning it around in a special pointer device, and can achieve a very sharp point.

The usual procedure is to lay in your lines with a hard lead (for example, a 3H). This is good for checking measurements and visually confirming the shape and form of a piece before making the final commitment to it. A hard lead leaves less graphite on the paper and is thus easier to erase cleanly. Having blocked in the main outlines and satisfied himself that they're correct, the designer proceeds to go over them again with a softer lead (I use an H) to get a darker line. At this point, he will also lay in the lines representing details of finish, such as moldings, hardware, and trim.

It is important to remember that the finished product of the drafting process is not your drawing, but the blueprint which will be made from it. It's a good idea to do a number of drawings on various weights of tracing paper and have prints made from them, just to find out how your particular drawing style will show up in the final blueprint. (*Blueprint* is a generic term; in reality, most such prints are properly called *blueline* or *blackline* since they are solid lines on a white background, and the process is properly called *diazo*.)

I spent many years in college getting criticisms for a thing called *line weight*. This will be discussed later, but it involves the thickness and darkness of a given line on the paper. After years of getting the line-weight mystique down cold, I was taken aback by a suggestion from a carpenter that I should make all the lines as dark as possible, so he could see them clearly through his protective goggles. Many people will espouse theories about drafting technique, but the only person whose opinion counts is the one who has to build from your drawings.

If you're working in ink, you'll skip the last stage and go directly from the hard-line lay-in to the inking-in step. Most drafting leads have a certain grease content and the softer leads seem to lay down more of it than the harder ones. Grease is anathema to ink (which is water based), so the less graphite you have smeared over the surface of your drafting paper, the easier will be the inking in.

The beauty of ink is that your line weights are strictly controlled by the point of your drafting pen and, since ink is more opaque than pencil, will print much more clearly. If you're going to work in ink, you have several options—you can buy a set of technical pens and a bottle of ink or you can buy disposable technical pens. Either solution is adequate for this work, and the latter is less hassle: the market is full of clog-free technical pens, but experience indicates that there is no such thing. Expect to spend a lot of time cleaning your expensive pen set, if that's the way you choose to go. After years of ecological responsibility, I gave up and bought disposables.

There is, of course, a third option now available: computer-assisted design (CAD) which refers to a whole spectrum of software packages that enable the operator to generate some fine and very accurate technical drawings when used in conjunction with a high-quality (and expensive) plotter. Unlike the graphic paintbox systems now available, the CAD systems were designed for the high end of the technical drawing market: engineers and architects.

This is not to say that they're not user-friendly systems, but that, in the same way that computer-assisted graphic systems remove a step between the artist and the final product (that is, the image on a TV screen) a

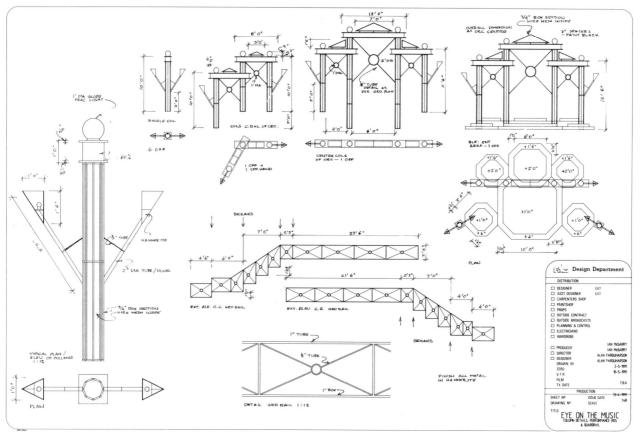

Figure 14.9 CAD construction elevations and details. RTE, design: Alan Farquharson

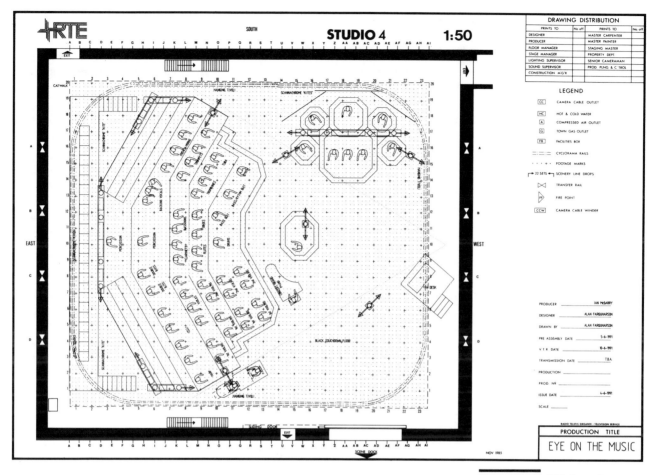

Figure 14.10 CAD floor plan. Computers are especially useful for reproducing complex items with minor changes, as is often done on floor plans. Design: Farquharson

CAD system used by a production designer adds a step of distance from the hand of the artist to the final product. Much of the work of set design is sculptural, and this is where such systems break down—the ability to draw free-form 3-D shapes quickly (and preserving your own style in the process) is often worth more than producing an immaculately clean computer-generated impression. Don't throw away your pencils and pens just yet!

To be fair, a CAD system can have certain advantages. The ability to construct images in three dimensions can be a useful way to visualize the studio set or location before actually committing to construction, although the more experienced designer will not need such a tool. The novice director who has impaired visualization skills will not gain much useful information from a computer- generated tour through the studio; the appearance of surfaces will not look the same, as the effect of light on an opaque surface is not easy to duplicate in a system that uses a cathode-ray tube (computer monitor) to display its images. For plotting set placement on the studio floor, and for the generating a certain kind of construction drawing, CADs are all right and often are faster. For instance in a show that repeats sets (such as a soap opera), the ability to generate daily floor plans quickly can be an advantage, and may justify the capital expense of the plotting equipment.

Language

I stated that floor plans and construction elevations are not realistic representations as much as they are charts or maps; as such, they have a set of symbols and conventions that amount to a sort of secret language, which you must learn in order to communicate via these drawings.

The mystique of *line weight* haunted me through much of my design training, because it seemed some sort of mystery of subtlety that was difficult to master. It is really not so difficult, nor such a great mystery.

There is a need to differentiate line weights in order to indicate the type of surface (or break in surface) is being described. The heaviest line indicates a cut through a solid surface, such as might be shown in a section view. The actual thickness of the line should approximate (in scale, of course) the thickness of the material being cut through: a section through a platform top built of ¾" plywood and framed with 1"x 6" pine should look like a slice through those materials, and be measured accordingly.

It is often standard practice to make the outlines of a piece slightly darker than the internal detail lines. This is good psychology; the heavier outline will tend to frame and define each piece and will assist the reader of the drawing in thinking of it in that way. As a rule, dimension lines should be lighter than the outline of a piece, so that they are not confused with other lines in the drawing. However, no lines on the drawing should be so light as to print weakly; this may well result in a misread dimension or an incorrect measurement from the print and cause a costly error.

When drafting site plans for location shooting, it is often useful to make the lines of the built dressing pieces slightly darker than the lines of the existing structures; this assists the carpenters in differentiating between them and may save errors in the field, where (believe me) they are much more troublesome.

It is also a good convention to do this on a lighting plot; if the lighting designer has the time (or an assistant) to draw the entire studio floor plan over again from the production designer's plan, it may be possible to lay in the outlines of the set pieces a bit lighter than the lighting equipment symbols and thus place the visual emphasis where it needs to be for the electrical crew. Since light plots often include a great deal of peripheral

Figure 14.11 CAD perspective drawing. CAD programs can simplify the difficult process of developing perspective views. Design: Alan Farquharson

information, such as color numbers, circuit numbers, and focusing notes, it's a good thing to reduce the indications of set elements to a light outline and thus avoid confusion.

If you're working with ink, the weight of a line is controlled simply, as a function of its thickness. By selecting a pen with a broader point, you automatically make the line heavier. Pencil is a bit more esoteric; the weight of a line is controlled both by the thickness and by the amount of pressure exerted on the pencil point. You can make a line broader by simply repeating it—in reality, laying down several lines so close together that they read as one. This requires a great deal more control and your state of mind (and emotion) comes into play. If the designer is upset or has just spent an hour lifting furniture, the likelihood is that the pencil will make incisions in the paper rather than lines; if the designer is weak from lack of sleep, the lines may be so light as to fail to print adequately. Pencil is a less accurate, but more expressive medium than ink—it's up to the individual to decide which aspect of the drafting is the more important one and choose the more appropriate tool.

Less subtle, and more critical to the accurate interpretation of the drawing, are the various symbols used to indicate specific sorts of construction techniques. There are not a great many to learn—at least not as many as are typically found in an architectural or an engineering drawing. Most of these are found in plan drawings, although some may also be used in others.

The most basic symbol is achieved by breaking up a line. There are a number of dotted or dashed lines in the usual vocabulary and each has a specific application. The simplest is a broken line of evenly spaced dashes, typically rather close together. This is a hidden line, indicating an edge that cannot be seen: It may be the line of the framing hidden under a platform top or under the facing of a flat. Such a line can also be used on a floor plan to indicate a line which is off the floor, but critical to the floor plan's use, such as a shelf or a header over a wall opening. Many designers regulate the length of the dashes fairly closely to indicate such different applications; typically, the short dashes indicate an edge of something hidden behind another part of the piece, the longer dashes indicate a line above floor level on the plan.

There is one special broken line, carried over from the proscenium theatre in which it is very important. This is the centerline (CL), which is used to indicate the center of the stage for the placement of scenery. It can

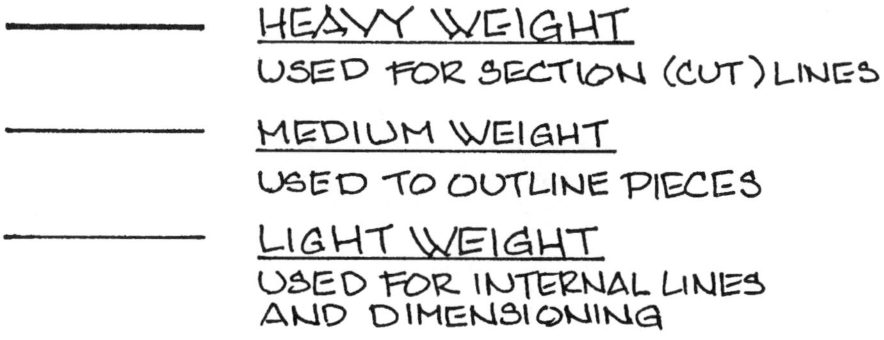

HEAVY WEIGHT
USED FOR SECTION (CUT) LINES

MEDIUM WEIGHT
USED TO OUTLINE PIECES

LIGHT WEIGHT
USED FOR INTERNAL LINES
AND DIMENSIONING

Figure 14.12 Line weights

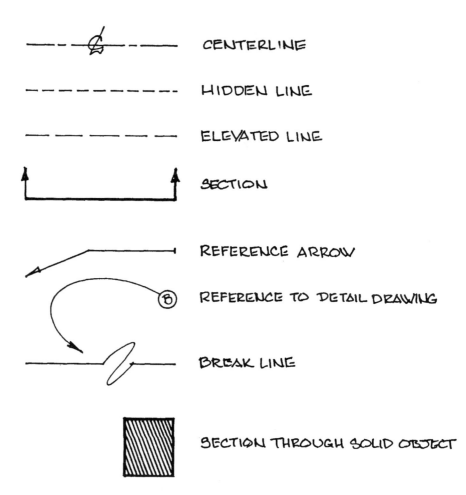

Figure 14.13 Drafting symbols

be used for the same purpose in the television studio, although the predominant grid system makes it unnecessary. Still, there are times when the centerline of a piece may be important and you can indicate this with a dot-dash line—a series of short and long dashes alternating with each other. Just to make sure you're understood, though, it's a good idea to label the line with a CL symbol.

There is a curious symbol, like an arrow turning a corner, consisting of a heavy right-angle line with an arrowhead on one end. This is the section line, indicating on the plan or elevation view the exact point at which the section view is cut and the direction in which we are looking at the section. This is often labeled with a capital letter, identifying the section drawing to which it refers, so there is no mistake.

Further arrows are often found connecting a circle with a letter in it to a detail on the drawing. This is a convenient way to make specific notes about finish of detailed parts; put your notes in a box and label each with a letter that corresponds to the appropriate one in the circle with the arrow. Brief notes on a simple and relatively uncluttered drawing may be made just next to the appropriate part of the drawing, with an arrow pointing to the exact location instead of resorting to the reference-letter system.

Another system of symbols and notations which is critical to understand, is dimensioning. Every construction drawing is done to a specific scale, expressed either in fractions of an inch to the foot (¼"=1'-0") or as a metric ratio (1:50).

Each critical dimension has its own dimension line to indicate its exact length. At the end of a dimension line is a short perpendicular end line, usually indicated with an arrow or an angled line through the point of intersection. The dimension line is usually broken at the center and the resultant gap is used for the measurement to be indicated. If it's feet and inches, the measurement will be indicated this way: 8'-10", indicating 8 feet and 10 inches. If the drawing is metric, the dimension will be written in millimeters: 1500 indicating one and one-half meters.

Dimension lines should be placed far enough away from the outline of the piece so as not to be confused with the drawing itself, but near enough for the relationship of the dimension line to its relevant points on the piece to be easily seen. As a rule (and for this reason), dimension lines for internal dimensions such as door openings should be kept outside the outline of the piece unless it's absolutely impossible to do so.

All construction drawings must include an identification(ID) block. This is usually located in the lower right corner of the sheet, so it can be easily seen when flipping through the corners of the stack of prints. Each drawing is assigned a number for easy identification. The ID block also includes a description of the piece (or pieces) shown on that page, together with the name of the production, the scene(s) in which it will be used, the name of the designer and the director, the name of the drawing's originator,

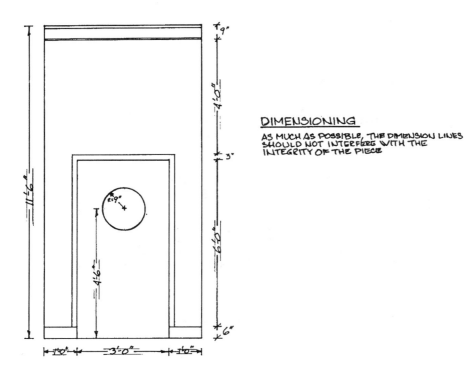

Figure 14.14 Dimensioning

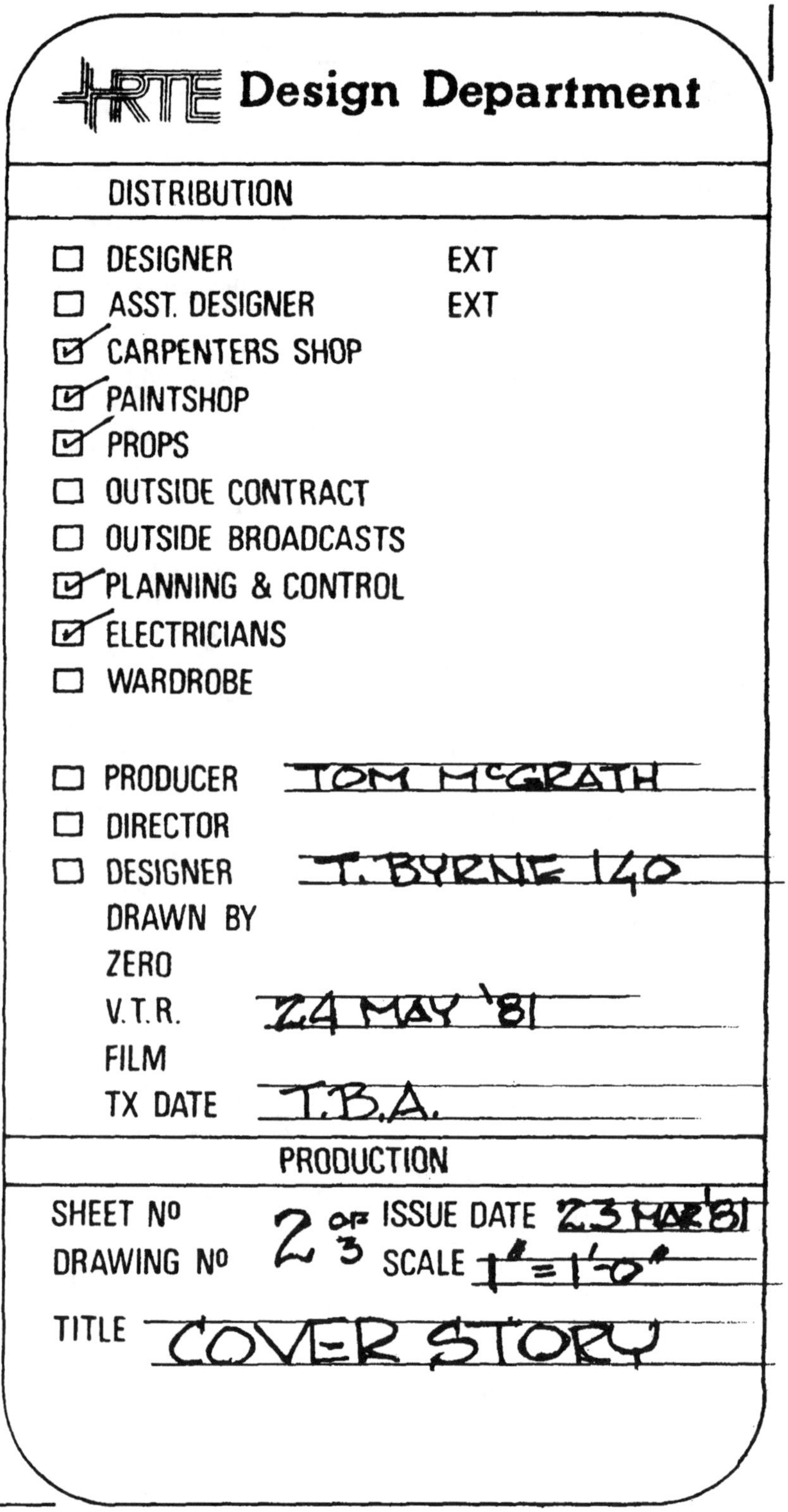

RTE Design Department

DISTRIBUTION

- ☐ DESIGNER   EXT
- ☐ ASST. DESIGNER EXT
- ☑ CARPENTERS SHOP
- ☑ PAINTSHOP
- ☑ PROPS
- ☐ OUTSIDE CONTRACT
- ☐ OUTSIDE BROADCASTS
- ☑ PLANNING & CONTROL
- ☑ ELECTRICIANS
- ☐ WARDROBE

- ☐ PRODUCER   TOM McGRATH
- ☐ DIRECTOR
- ☐ DESIGNER   T. BYRNE 140
- DRAWN BY
- ZERO
- V.T.R.   24 MAY '81
- FILM
- TX DATE   T.B.A.

PRODUCTION

SHEET Nº 2 OF 3   ISSUE DATE 23 MAR 81
DRAWING Nº 2 OF 3   SCALE 1"=1'-0"
TITLE   COVER STORY

Figure 14.15 An ID block

the departments to which it must be distributed, and the *zero date* (the deadline for completion). It must also include the scale to which the drawing is done, even if this is also indicated elsewhere. In a larger institution, there will also be spaces for various managers to sign their approval and indications of whether the scene is to be filmed or taped, whether the piece will be used in the studio or on location, and whether the piece is to be used once and recycled or kept for future shoots.

Style

I've noted that a good construction drawing conveys a sense of the designer's artistic flair as well as the essential information. The way you handle the media you choose will affect your drafting style greatly.

Some designers are more precise than others and some impart a stronger sense of the finish of a piece than others. Usually these two qualities, accuracy and ambiance, are at opposite ends of the continuum.

Those who prefer pencil tend to fall into the artistic group and those who prefer pen or CAD tend to draft more like engineers. I was taught to work in pencil and to weight the lines more heavily at their ends than in the middle, as well as overrunning slightly at the corners. The theory is that this draws attention to the points at which lines cross—the natural reference points of the drawing—and thus imparts more importance to them in the mind of the person reading the drawing. I also tend to indicate details of molding and trim fairly accurately and even shade in parts for a more 3-D look when I think it's appropriate.

Other people may prefer to draw with a pen and give more accuracy to their line weights and make a generally neater drawing. These people also tend to leave out indications of finish on the construction elevations, saving that for the painters' elevation. Some people like to draw large, and will work in ½" scale. I prefer this because I can rely on the carpenters to take more accurate measurements from the drawing should I omit some in the dimensioning. A famous designer, Donald Oenslager, used to do elevations at ⅛"=1'-0", and they were accurate! (I can imagine, though, that he may have caused the blindness of several construction chargehands.)

I try to organize the different views of a given piece in a logical way: I will arrange them around the elevation, putting a plan view below it and a section view next to it. This makes measuring easier for me as I lay in the drawing, and shows the carpenters the relationships of the ancillary drawings to the main view. It is sometimes necessary to show an awkwardly shaped piece in a three-dimensional view. There is a device called an oblique projection, which shows a three-dimensional piece as seen from an angle, but with all sides shown in scale so they can be measured. This is alright as far as it goes; the net result is a distorted view, since it seems to defy the laws of perspective. I prefer to do a sketch of the finished piece from an angled view, showing its 3-D qualities, and labeled "not to scale" next to as many flat dimensioned side views as are required to show every measurable edge. These views may be supplemented with many sections through the piece from many different angles, to show the internal framing

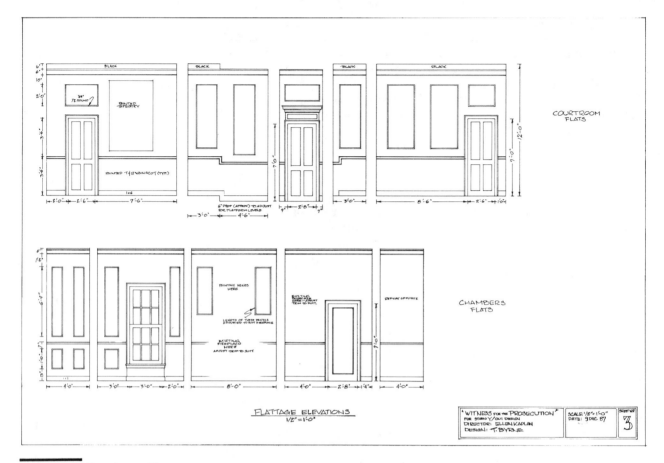

Figure 14.16 Page layout. Note that sufficient white space has been left for shop workers' written comments and sketches.

as well as the relationship of the outer surfaces. I also try to leave a fair amount of white space; this tends to seem less overwhelming if the piece is complex, and leaves room for me to make supplementary sketches when I discuss the construction details with the chargehand carpenter.

The trick with construction drawings and studio plots and plans is to strike a balance between expressing your own design style and communicating accurately the details needed to build faithfully. The best way to do this, I find, is to keep in mind the person for whom you're making the drawing—the one who has to interpret it in the execution phase of the process. You want to communicate well with that person, both the style you want for the finished piece and the basic numerical and technical information necessary to execute it with accuracy.

Sets and Props   15

Working Conditions

It is one of the sad facts of a shrinking economy that the major entities in the television industry no longer maintain their own in-house construction facilities. The networks' commitment to their own productions no longer extends to the maintenance of a salaried staff of carpenters, painters, and prop builders. This is due almost entirely to bottom-line management brought about both by dwindling revenues and the need to make bigger profits and support a larger staff at the senior management level. Production designers also are now being hired on contract, varying from yearly to per-diem or even hourly rates of pay.

You may well imagine that this system has an impact on the working life of the production designers, apart from the fact that they, too, are usually thrown into the contract-labor pool without any guarantee of employment continuity or benefits. In the old days, the designer was the pivotal communicator between the director and the construction personnel; in the present system, that role has simply intensified. Whereas the carpentry shop used to be part of the same company and often was located in the same building as the production studios, it is now somewhere else and has no specific commitment to any one client's priorities. (Often the same shop is not used by a studio for any two projects.) If, as is often the case, your director makes some rushed last-minute changes, it is a source of substantial angst for all concerned: The shop may have another project in progress and be unable to afford the time for changes, or you may incur substantial amounts of overtime pay for something that may not have been a serious problem under the old system. Of course, it is you, the designer, who will be expected to rush from one place to another to convey information and hurry things along. The result now is that the designer's life has gotten very hectic indeed, moving from his studio to the director's office to the construction shop (or shops) to the production studios. In his spare time, the production designer is meant to come up with some dynamite creative solutions.

These are now facts of life; the networks are run by business people, not television people, and they see spreadsheets where you or I might see a storyboard. If you're anticipating being a designer at this level, get used to the idea that you'll spend most of your working day on your feet and that you won't have a lot of time off, except for the months of downtime when you have no work at all!

As a result, the position of the production designer becomes much more pivotal and more of a business management role than a creative one. Your drawings will become contract documents in the strictest sense and the exactitude with which you draw them will bear a direct relationship to cost accounting and budget targets. You must also concern yourself about manhours, transport, storage, and maintenance costs, and have the travel time between stations (railroad stations) nailed down to a T.

Building

During the early years of television, there was some competition between the two traditions of stage carpentry—theatre and film—to see which would prevail in the new industry. As Frank Schneider put it, sets built in Manhattan were built in the theatrical shops and used theatre techniques; those built in Brooklyn were done film style.

The truth is that film-style construction won overall since it is really better adapted to TV. The old theatre *flats* were made of canvas or muslin stretched on a frame of 1" × 3" pine and were slightly floppy when the weather became humid. This is acceptable when your audience is fifty feet away; some slack canvas will not really destroy the illusion. When you're shooting close-in, however, this becomes a major difficulty.

Other elements of detail also became important. It was no longer sufficient to paint detail in the *trompe l'oeil* fashion that was acceptable in

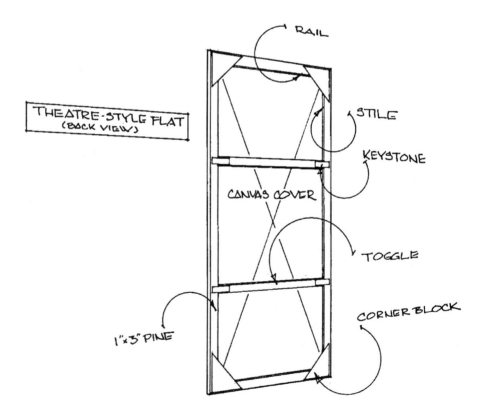

Figure 15.1 Flat construction: theatre style

the theatre. It was also no longer necessary to make things very strong and durable; a TV set, like a film set, often needed to be erected only once to fulfill its function.

Still, the background of the system was theatre and most of the terminology is the same. We'll look at each type of scenic unit individually, taking into account the different ways of constructing and finishing each.

Flats

A *flat* is a lightweight vertical unit, usually used to represent a wall. It can be finished in any of a thousand ways to represent any of the types of walls found inside or outside any building. Flats can also represent less regular surfaces, such as cave walls or cliff faces, by using some rather radical surface treatments.

The typical TV flat is known as a *hard flat* because since the surface covering is plywood rather than canvas. Apart from not flapping in the breeze, the hard flat also provides an infinite number of places to hang things such as pictures, clocks, or sconces. It is slightly heavier than a canvas flat, but since TV sets tend to be less tall than theatre sets, hard-faced flats rarely become so heavy as to be cumbersome.

Trompe l'Oeil is painting to simulate 3-D reality.

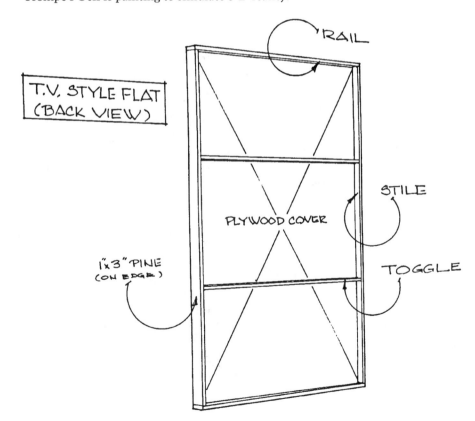

Figure 15.2 Flat construction: TV style

Hard flats are typically built with a ¼" (3mm) thick sheet of plywood. This is then framed for support on the back (off-camera) side with 1" × 2" or 1" × 3" pine stood on edge (attached perpendicularly to the plywood) for maximum resistance to flexing and twisting. The plywood facing may be attached to the frame by nailing or screwing. Is important that the surface be marred as little as possible; to this end, the best method is to use finish nails and glue the joint for additional strength. Sometimes it may be more efficient to use drywall screws, especially when the flat is of an unusual shape and will not likely be saved for reuse. These screws can be driven very quickly indeed with special drills, although they tend to split the framing stock if not driven carefully.

Even when supplied in metric thicknesses, plywood still tends to come 4 feet wide by 8 feet long, so it still is important that we design with this modular unit in mind. Eight feet is often not tall enough for studio use, and a 10- or 12- foot flat will be more typical. In such a case, remember to provide framing members at the joint between the pieces of plywood. Often, you can contrive to have a picture molding or some such detail at this point, to help cover such a joint, although there are other, less obvious, ways to hide blemishes.

A flat almost never stands alone. Typically, they may be used in runs consisting of several flats joined together to make a single long wall. They can be joined to each other by screws through the framing members, or simply clamped together with standard carpenters' C clamps (sometimes called G clamps in England). A wall usually has some additional means of support, called braces or jacks. The old theatrical system used a *stage brace*, which was a wooden pole, in two telescoping pieces, with special fittings on the ends to attach to hardware on the flats and on the floor. This is fine, except that it turns the floor to splinters. The floor end of the brace can be attached instead to a plate with weights on it, but this can slip under extreme pressure and allow the wall to collapse. A better (and cheaper) solution is to build your own supports called *jacks* (or *French braces*). Each of these is simply a right triangle made of 1 × 3, attached with backflap hinges to the flat. When the flat is in place, the jack is simply swung out and loaded with iron weights or sandbags to keep the whole unit from shifting. This system works well and does little or no damage to the studio floor.

When you need a wild wall to be easily removed for a camera shot, you can use a device known as a *tip-jack*, which allows the flat to be tipped backward onto casters and rolled out of the way until it is needed again. The film industry has developed steel units that clamp onto the flat at the sides and roll it away. These take a bit longer to fix onto the flat, but work well and roll across the studio floor a bit more smoothly, because the load is better balanced. Of course, if the flat is small, a couple of stagehands can simply carry it away. The simplest solution is always the most efficient.

The job of set construction would be very simple if there were nothing but plain flats to build. However, most walls come with openings of various sorts and appendages that complicate things.

Obviously, few walls do not have doors in them. The door itself may be built in the shop or bought from a lumber dealer ready-made. Of the two types of doors, panel and hollow-core, the older style is the panel door and

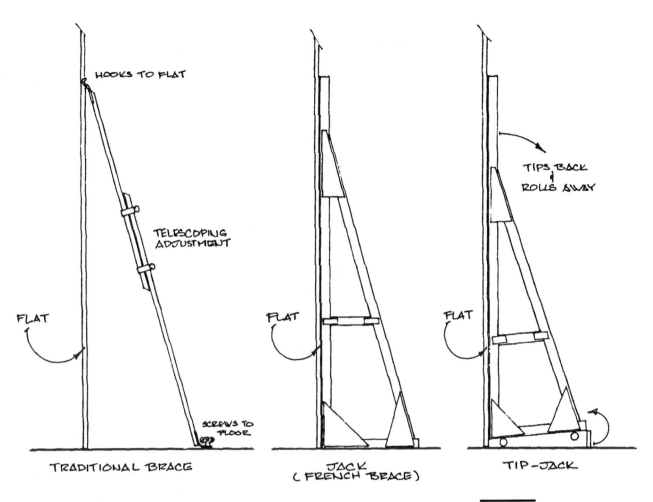

Figure 15.3 A jack and a stage brace

it tends to be heavier. The hollow-core door is the more modern and is lighter, being constructed of thin plywood veneers over a frame of pine and a cardboard honeycomb filler. Either type may be useful, although the heavier door usually requires additional reinforcement of the flats to resist the shock of its impact. The door frame is usually made of 1" × 6" pine, set on edge to the surface of the flat, and attached to the framing. This material is referred to either as the *thickness* or the *reveal* depending on colloquial usage and the inclination of the user toward pragmatism or mysticism. The door frame may be built into the flat, in which case it is known as a *dependent* door; or it may stand on its own and be inserted into the door opening in the flat, in which case it is called an *independent* door.

The door frame must be reinforced to resist the twisting force exerted on it by the swinging of the door itself. This is best accomplished with small blocks of wood or angle brackets. Each door frame must incorporate a stop-strip around the top and sides to make sure that the door stops when closing and to keep it aligned with the latch hardware. I prefer to mount

Figure 15.4 A wild wall at Warner Studios, Burbank

the door inside the frame in this manner rather than attaching it to the back because it looks like the real thing in a close shot. The latch and lock hardware used is the real article, bought from your local hardware store and mounted as it would be in a real building. (Remember to find out if the door needs to have a working lock and to make sure that your props crewhead has the key on the shooting day!) Across the bottom of the door opening will be a wooden sill, the long edges tapered (beveled) so as to avoid tripping anyone who walks through the door.

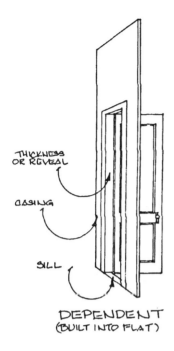 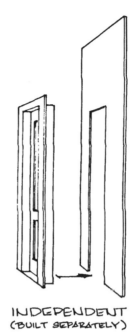

Figure 15.5 Doors

Walls also commonly have windows, and the general technique of construction is the same as it is for doorways. The opening will be framed with 1" × 6" pine, which may be dependently or independently made. (It is more common to make windows dependent, because most are much lighter than door units.) Window units can be bought factory-assembled and ready to pop into a flat; these are available in a wide variety of sizes and may actually save money in some cases.

Several types of windows are available, and each has its own method of construction. The most common window is the sash-type, which has two sections that can slide vertically within tracks in the frame. Strips of wood like those used to make door stops are attached to the sides and top of the window frame to make two tracks. The sashes slide up and down in these tracks, the top most one being on the outside of the building and the lower one on the inside.(In real architecture, this ensures us that the rain will flow away from the building rather than pour inside.) Each sash may be a single pane of glass or it may be subdivided into smaller panes; typically, we use a single panel of plexiglass (perspex) for each and then dummy-up smaller panes by gluing molding pieces onto the plastic. If you want a stained-glass effect, there is nothing better than to stick pieces of lighting gels onto the clear plastic.

The other type of window is called a *casement*. It operates much like a door, swinging out on hinges away from the building. Casement windows are typical for older buildings and castles, for example. There will usually be a solid piece mounted vertically at the center of the opening to which the casement window panels latch when closed.

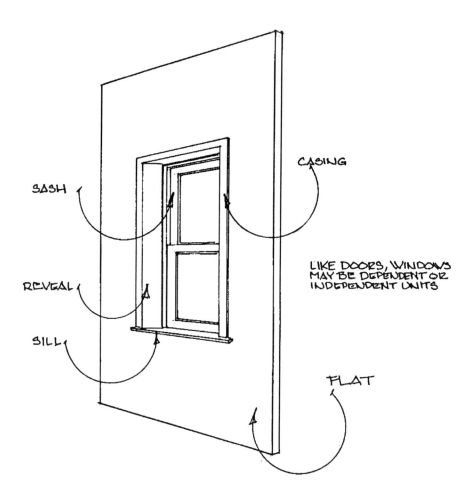

Figure 15.6 Windows

There are a number of detail pieces that may be mounted on the face of the flat. Doors and windows will typically have a trim of some sort of molding around the top and sides of the opening, which is called a *casing* and can take any of a number of shapes depending on the period of the piece. A window will also have a sill on its bottom edge, which will have a decorative molding beneath it.

Along the top of a wall there may be a *cornice* molding, which may vary from a single smallish piece to a downright huge construction, again depending on the period of architecture being represented. When a massive construction is required, the normal practice is to make a box with a number of smaller moldings attached one above the other for the desired effect. This is supported with several profile pieces which are attached to the flat either directly (by nails or screws) or by hanging brackets. There is a whole language of cornice molding terminology, incorporating words like *dentil*, egg and dart, and ogee, each of which describes a different shape. Study your architecture!

Other surface moldings may be *chair rails*, *picture rails*, or *baseboards* (skirtings). The classical styles of wall treatments may include

Sets and Props 133

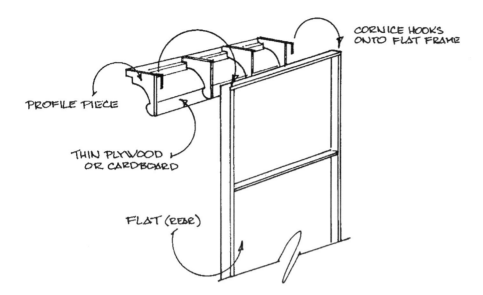

Figure 15.7 Cornice

fake columns (called *pilasters*) or *panel moldings*. These can often be used to hide joints between flats, as can slight changes in direction of wall sections, called *jogs*. Treatments like these can be very useful in planning wild-wall sections.

Platforms

Eventually, you will design a show that needs levels. This is director-speak for platforming (rostra), and refers to the visual variety which can be achieved by raising the floor level at certain points on the set. Often, this will occur as an architectural feature of a drama set, such as a stair landing or balcony. Even more frequently, variety shows will need constructions involving many levels of platforms, without any need for architectural justification. Dance numbers, especially, can make good visual use of a set with many levels.

Just as there have evolved several different ways of building flats, so have a number of platform configurations come into use over the years. The oldest one still in use is the parallel platform. This is most useful in a situation which requires that the set be *struck* and re-erected many times, especially where storage space is limited. The idea is that each platform has a set of supporting legs (really a sort of truss), which fold almost flat, the side sections remaining parallel to each other, whether folded or open—hence the name. This truss system has two long side sections and two short end sections, all of which are joined by a rather complicated system of hinges. There is a third section, supporting the middle, which may be removable or may be permanently attached, depending on the type of parallel being made. The American style has all three cross pieces attached to the side frames, the continental a removable center section. In the parallel platform (rostrum) the

---

To *strike* a set is to remove it form the studio.

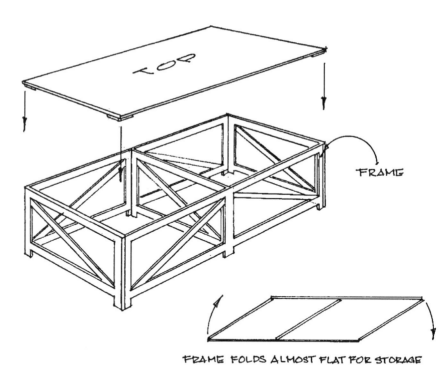

Figure 15.8 Parallel platform. Note that the support structure is really a parallelogram when it is folded for storage.

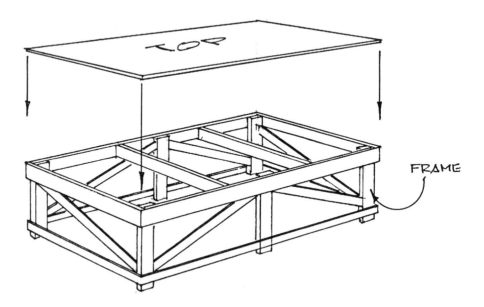

Figure 15.9 Rigid platform. These are framed with 2 x 4, ⁵/₄ x 6, or even 1 x 6

top is not framed; it is usually just a sheet of ¾" plywood fitted with blocks of 1× 3 to prevent it from shifting when the unit is assembled.

The other predominant type of platform is that which uses a framed top. This, again, consists of a sheet of ¾" plywood, framed to be self-supporting. The framing may be 1" × 6" pine set on edge, or 2 × 4 stud timber. Since resistance to flexing is a function of the depth of the supporting member, the 1 × 6 system is both more rigid and lighter, and may, therefore, be preferable.

This system requires separate legs, cut to specification. These may be 2 × 4, 1 × 6, or whatever is strong enough; the choice will be made based on the height of the platform. If it's very high (more than a few feet), it is wisest to include a system of triangular bracing between the legs and the platform framing to prevent the platform from toppling.

There are other details of platform construction that you will need to know. Platforms are notoriously noisy constructions and a number of techniques are available to help with this. The usual culprit is impact noise caused by the shoes of the performers coming down hard on the platform top and making each step sound something like a bass drum. The old (and still workable) solution to this problem was to cover the platform top with carpet felt and then cover that with a layer of very heavy canvas, which could be painted. This is a good solution acoustically, but it tends to round off the corners of the platform tops, and make them look less realistic. The tops can simply be covered with canvas; this will lessen (but not eliminate) the impact, but will not alter the shape of the edge. Some people add a layer of soft composition board (such as Homasote) and cover that with a thin sheet of hard surfacing (such as Masonite). This gives the realistic click of shoe contact without the annoying drum resonance. The final alteration is to attach carpet felt to the *underside* of the platform top; this dampens the resonance between the platform top and the studio floor, and further reduces the drum effect.

Stairs

There is another level that always goes along with platforming: the stair unit. Just as a wall flat is of little use without a door, a platform is of little use without a stair. And like the door unit, a stair unit may be described as dependent or independent. As you may guess, the independent stair is one that will stand on its own and the dependent stair is one that attaches to a platform for support. In cases of space limitation, the dependent is preferable, since it stores more compactly.

As the designer of a platform, you are partially liable (along with the scenic shops) for any damages from injuries, should your platform fall. Because of this, many scenery shops tend to overbuild and over-reinforce. This is not a bad thing; people have been killed in stage and studio accidents. As the manager of the design budget, you are likely to be tempted at one point or another to cut corners on construction. Please don't ever compromise the safety of your performers and crew by saving money on safety features!

Any stair unit is comprised of three basic elements: the *tread* which is the part you tread on; the *riser*, which is the vertical part of each step; and the *stringer*, which is the side member that holds the whole staircase together. The tread is made of ¾" plywood or 1" pine, and is usually from 10" to 12" deep. The riser is also usually ¾" plywood or 1" pine stock, and may be 6" to 9" high. (It is a bad mistake to make your risers unusually high or low; people are accustomed in normal use to risers within these parameters and may likely trip on anything else.) Having decided on your tread and riser dimensions, you may now go ahead and work out the design of the stringers. If the unit is to be independent, these will likely be large plywood triangles with the stair line cut out, sawtooth fashion, to fit the profile of the steps. If dependent, the stringer swill be 2 × 8 or 2 × 10 pine, again having the step profile cut out of one edge. The independent unit will be framed internally to be self-supporting and (unless it happens to be very small,) will typically have a system of casters to permit it to be rolled around the studio.

Stair units also commonly have handrails attached. Both for reasons of architectural realism and (where the stair is not seen) for reasons of safety, a handrail is a good idea. In situations where a handrail is not an option, such as an open stair on a variety program set, it is better to keep the rise small (6 to 8 inches) and the tread generous (12 inches or more).

Figure 15.10 A step unit

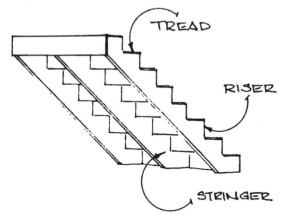
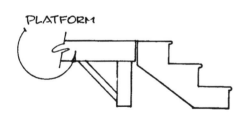
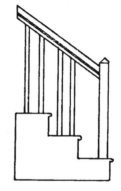

As with other architectural units, stairs have their own special details and terms. Most household stairs have a half-round molding on the front edge of each tread, called *nosing* The spindles that support the handrail are called *balusters*, and a row of them a *balustrade*. The big posts at the top and bottom of a balustrade are the *newel posts* and the platform at the top of the stairs or at the middle (for instance, when a stairway changes direction) is called the landing.

Apart from being excellent blocking devices for directors, stairs are wonderful decorative devices and give the designer a great opportunity to do some sculptural work and display her command of detail and period finish.

Special Construction

Thus far, I've discussed only wooden construction techniques. These are the tried-and-true methods and those on which the more exotic methods are based. Nonetheless, with the spiraling rise in timber costs, metal and plastics have come into their own in a very big way.

The principal use of metals in scenic construction is, of course, structural. Especially when a construction is large and heavy, the use of steel or aluminum framing makes a great deal of sense. Metal structures also stand up better in the weather and therefore, are often used for permanent exterior sets, such as the one used at Elstree Studios in London to shoot *Eastenders*.

The flats and platforms are typically constructed in the usual way, with timber. If the piece is to be used outdoors, the finish will be waterproofed onsite before assembly. The metal parts would then be the legs and other members of the support structure.

If the set is to be used in the studio, there is often another compelling reason to use metal: scaffolding. There is a wide array of quick-assembly scaffolding units available commercially, and they make great underpinnings for large set units. They also transport and go together quickly and easily, and can be rented from certain vendors, making them a cheap alternative. Again, these would be used to support a set of platform tops built in the customary manner.

Plastic is another material that has come into its own in the last 30 years. There are many types of plastic available and each has specific properties that make it useful for specific applications.

While most plastics are used as finish treatments, to achieve sculptural effects cheaply and without excess weight, some are also useful in structural ways. Some technicians have advocated the use of rigid plastic foams as core materials for the middle layer of a sandwich of foam and two outer layers of a harder substance, such as plywood—rather like the foamcore board used to make models. I've never seen such an application and I imagine it's gone from our catalogue of options.

The most structurally useful plastic is probably fiberglass. This is a glass-strand fabric sealed in a generous soaking of polyester resin. The polyester is painted onto the glass fabric and left to harden. When it has thoroughly hardened, it is amazingly strong and resilient. I've often made tree trunks from two pieces of ½" plywood and a single vertical 2 × 4: I just

Figure 15.11 The backlot at BBC's Elstree Studios: *Eastenders*. Note the steel framing and general permanence of the construction.

staple glass fabric around this structure to make the sides of the treetrunk and then give the whole thing several coats of polyester resin. I know of three of these that withstood several years of abuse as props for a dance company and that are probably still around.

There are many kinds of sheet plastics, which I'll discuss in a later section (3-D Treatments). One of these, though, is polyvinyl chloride (PVC), which comes in a wide variety of tube sizes, and may have some limited structural applications. Acrylic (plexiglass or perspex), too, is commonly used in place of glass in sets, because it is easy to cut and safer than real glass.

Furniture and Props

When luck is with you, you'll walk into the prop storeroom and find exactly the furniture you have in mind just sitting there waiting for you. Such luck is rare.

Compromise is the name of this game, although you should make every effort to minimize it. Normally, having sifted through and digested the contents of a number of reference books, magazines, or other material, you'll have some sketches of the furniture you want and you will set out on your quest. If your budget is high, you may go to a furniture rental house. This is the luxury end of the market; some of the will alter or even reupholster pieces for you (for the right price). Most of us go instead to the junk shops and sort through mountains of dusty, moldy castoffs looking for something that will do.

These pieces are then carted back to the props workshop and cleaned up or altered as necessary. Upholstery is a highly developed craft and would

Figure 15.12 Construction with bamboo: "China Beach". Carpenters building a thatched hut on location. Design: Joe Wood.

fill another book. The repair and modification of furniture generally is also a highly skilled craft and best left to those who know how. It is true, however, that television is a forgiving medium: Because the camera actually leaves out much detail from props (leaving it to the brain to fill in missing information), small nicks and scratches often don't show up on camera. This is a good thing, considering the abuse prop furniture gets between shoots. (A designer friend of mine, Mick Grogan, who saw a demonstration of the 1125 high-definition system at a trade conference in 1982 and could only say, "We're ruined." Everything showed up in shot in the clearest detail—especially the flaws.) A great deal can be done to a fairly plain piece of furniture to make it look more like it ought. Given the forgiving quality of present-day video, we can make rather crude alterations which look fine in shot. Fancy scrollwork can be applied to a plain chair by gluing on plastic leaves from fake plants or by carving foam or building up papier-mâché, and painting them to match. Various stock bits of detailing, such as turned or shaped newel posts, rosettes, or fancy moldings, can be applied to cheap pieces of furniture from the junk shop to make them appear to have come from Versailles, or at least a better class of junk shop.

Sometimes, of course, you can't find anything even remotely resembling what you need, and the only answer is to build the piece from scratch. This is not too difficult if you have a creative inclination. A good hand with a sabre saw and a router, and a source of the already mentioned moldings and turnings can yield a Renaissance table in a matter of a few hours.

Finishing

Whether the work is done in-house or jobbed out, there will usually be a staff of painters in the same facility as the carpenters. This is just good sense; it's much more efficient to work this way and, even though the painters are often freelancers, they usually work in a paint area of the construction shop. Typically, they are responsible for any interpretative decisions on the construction floor. Often, the scenic artist will be asked to lay out intricate shapes for the carpenters to cut and to do any sculptural work.

3-D Treatments

Much of the work done with plastics may fall to the scenic artists. The most common use of plastic foam is sculptural, and it is ideal for this use. There are several types of rigid plastic foams available, which are really designed for use as insulation in buildings. Most of these are now made flame resistant, although they tend to give off noxious fumes when burned or melted.

Any textured wall surface can be achieved with rigid foam. Stone, brick, concrete, rough wood, and steel girders can be quite effectively mimicked in foam, and with an immense saving in weight! Foam can be cut with a saw or with a hot resistance wire and can be textured with various solvents to achieve a weathered look. It can be glued to a hard flat or other supporting structure; it needs a rigid support to stand. It also needs some kind of skin to take paint and to protect it from impact. I find that a skin of muslin scraps glued onto the foam after it is carved provides a superb sur-

Figure 15.13 Construction with plastic foam. The details being built into this pub set are cast plastic foam. RTE, Design: Lona Moran

face for paint and is remarkably durable. Apply the skin and glue it with thin polyvinyl carpenter's glue and a paintbrush; do this while the flat is lying on the floor, so gravity will help your muslin follow the contours of the carving.

A number of specialist manufacturers sell a wide range of vacuum-formed plastic sheets for detailing flats. These can be rosettes, door panels, fancy carvings, or even whole bookshelves (with books). Used carefully and judiciously, these can be made to look very convincing. They're very cheap compared to other solutions, although they are a bit fragile and often need maintenance work. In its 14th Street studios in Brooklyn (formerly Warner Brothers'), NBC has a scenery shop with a 6-foot vacuum press for the manufacture of these panels. They no longer build in-house, so it stands idle, but with such a machine at hand, one can design and make molds from which to mass produce architectural detailing out of styrene sheets. A number of books are available that explain the procedure for construction and use of a vacuum press, if you're so inclined.

Painted Treatments

Having achieved the correct shape and texture, the final finish step will be to add a paint treatment. Except in cases where a manufactured surface treatment, such as wallpaper or fabric, is used, the finish is almost always a painting technique.

Scene painting is an ancient craft and one requiring a high degree of skill and talent. Many people call themselves scenic artists; few really are. Nolan Brothers' Studios, in Brooklyn, is one of the remaining bastions of the scenic artists' craft, and they turn out some remarkable works of artistry. The BBC has a fine scene painting department and makes good use of the talent there. Otherwise, it is a disappearing trade, and much of what is deemed acceptable scene painting for broadcast demonstrates this.

Traditionally, scene paint was a sort of fresco medium and was literally manufactured by the scenic artists or their assistants. This involved

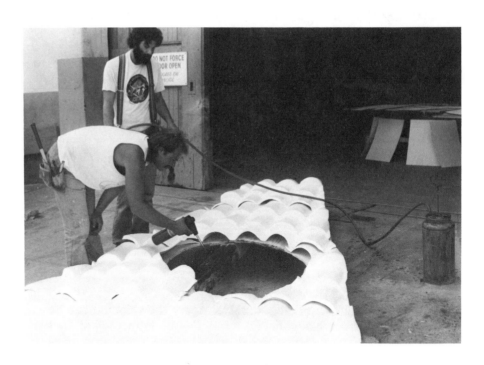

Figure 15.14 Scenic artists building a tile roof. The tiles are vacuum-formed plastic sheets.

mixing dry colors with water and ground-flake or casein glue to make paint. All that is essential for making paint are a *pigment* (for color), a *vehicle* (or solvent), and a *binder* (glue). The pigment contains the color properties, the vehicle gets the color onto the desired surface, and the binder keeps it there.

Beautiful in its simplicity, the old scene paint system is still in use (I use it) and has distinct advantages: The colors are the most brilliant you can get, and the system of mixing them is quite accurate, since other systems tend to change the color a lot from wet to dry states. Of course, it has disadvantages: The binder rots and stinks, some colors bleach out in the sunlight, and it isn't waterproof. The first problem can be solved by adding some antibacterial agent (Lysol) to retard the rotting. The other two can't be overcome.

Acrylic or latex (or emulsion) paint, which is water based, uses acrylic plastic as the binder. (This is the kind of paint you use on the interior walls of a house.) You can, of course, get it at the local paint or hardware store, but the colors prepared for home use are usually paler tints than most scenic artists prefer. This is expensive and inefficient; buying household paint means that you pay top price for a lot of water and white filler, when it's the color that you really want. There are proprietary acrylic systems available to permit you to mix your own and these are available to the paint merchants. Typically, this involves buying the liquid binder in a concentrated form and adding concentrated liquid pigments, then letting down with white pigment filler and thinning with water to the desired consistency. It's a good system and one that gives a waterproof finish, although it takes more practice (and usually a few test patches) to get exactly the right colors. It's also bad for your scene-painting brushes: The acrylic binder will not wash out of the bristles if allowed to dry. (A set of good scenic artists' Chinese-bristle lining brushes may cost several hundred dollars.)

Another water-based medium, which only works well on cotton canvas or muslin, is aniline dye. For background cloths and such, aniline dyes are superb; they have a luminance and saturation that nothing else can touch. A particularly spectacular effect can be achieved by painting a scene on canvas with dye and lighting through the cloth from behind.

Anilines are not for the faint hearted, though. There is no room for error here; if your hand slips on a critical detail, you may ruin the whole piece. Dye reacts with the fabric and becomes part of it in a way that paint does not. If you're going to work with anilines, you need to be a very practiced and confident scene painter. Still, the effects to be achieved with this medium are impossible with any others.

Don't ever use aniline dyes outdoors; I once painted a large canvas show curtain for the musical *Sugar* outdoors on a sunny day. After our lunch break, my crew and I returned to find that all the reds had bleached in the sunlight—wherever we'd painted pink, it was white again! Since roughly a third of the curtain was shades of pink, this was a disaster.

Though the water-based media will suffice for most applications, there may be occasions when oil-based paints may be necessary. Location uses, which may need to withstand weather over long periods of time may require an oil paint, although there are now some very good weather resistant acrylics.

There is one very useful technique that takes advantage of the fact that oil and water don't mix. This is called a *French enamel varnish* (FEV)even though it isn't French and doesn't use either enamel or varnish. The method is to simulate a wood grain with a brushed finish by painting a light-toned undercoat with a water-based paint, then brushing the grain effect over that with shellac and a darker shade of aniline dye. The effect is quite like wood, once you get the knack of the brushwork, and has a nice luminance and depth that look very good on camera. As with other waterproof finishes, shellacs, lacquers, and oil paints tend to clog the bristles of the brush unless thoroughly and quickly cleaned after each use.

The surface treatment of the piece, whether set piece or prop, makes a huge difference to the finish. You need to know before beginning the type of surface you want: should it be dull or reflective, should it have a coarse or a fine texture?

Scene paints and dyes take very well on fabrics and not so well on smooth surfaces or on nonabsorptive ones. If you want to paint a piece with either of these media, cover it with a skin of muslin before you paint. Wall flats look much better under a paint treatment if they're skinned with muslin first; it gives a depth and brilliance to the color, makes the paint flow more easily (which is good for painting details), and is much less likely to throw a reflective glare back into the camera.

Acrylics and synthetic paints do less well on fabric, but cover raw wood well. FEV techniques look awful on fabric; only a smooth surface such as raw wood, will do for these. The same is true for oil-based paints. As a rule, cloth-skinned flats are good for studio use only, because they don't stand up to moisture well.

Plastic surfaces can be a real problem. Many of the water-based paints will not stick to them, and oil-based paints may actually cause the plastics to dissolve. The definitive reference work for this is *The Intergalactic Serial Shop Cookbook* by Ned Bowman. This book contains charts relating certain plastics to their compatible finishes; it's a complicated subject, because there are so many different types of plastics available.

As a production designer, you should acquire a library of such reference works. While most scenic shops have talented and knowledgeable people on staff, it's best to know before you begin the construction exactly how a certain piece can be built and a certain finish achieved.

Set Dressing

The final step in finishing a set is dressing it. I've mentioned dressing already; it's the process of choosing and arranging the items of detail that add the final touch to the look of the set. In the case of a realistic drama, these would be the items of room decor and personal effects of the people who are supposed to inhabit the space. In other contexts, the dressing pieces may simply be there to add texture or atmosphere. Whatever the format of the program, these details will make the subtler interpretative statements.

Whether just a vase of flowers or a shelf full of books, the dressing props will help you to make a visual comment on the program or on a particular scene. The design choices made for the set will tend to be broader ones because the set is designed earlier in the sequence of events and must do for all of the scenes. Dressings are smaller, cheaper, and more portable; they can therefore, be changed with greater ease and speed, can be used to put a specific spin on a specific scene, and then can be removed for the next scene. Also, important dressing props can be used in the foreground of the shot and thus given greater importance in the eyes of the audience—greater even than the actors in some cases.

The process of dressing is part of the propping process and involves the same skills and drudgeries. As you're searching through the secondhand shops for furniture, you will also have the list of dressing props in case you run across something useful while looking for something else. (This is usually the case.) The more difficult dressing pieces will fix themselves in your memory, so that, on the way to shop for groceries or something else, you will suddenly stop and run into a shop you've never noticed before to buy a whatsit you saw in the window, which is just what you needed to put on a shelf in shot in a particular scene. People think production designers are strange; it's an occupational hazard.

Once the dressing pieces are assembled in the props room, the designer and the set dresser or props head will sit down together and carefully catalogue all of them. This will result in a chart or a book of charts, cross-referencing the item (by name or number) with the scene and the set or location in which it is to be used.

The dressings for each set are then carefully stored on their own shelves in the props room and hauled out to be placed on the set for the initial shoot. The designer will direct this process, making sure that each

Figure 15.15 Set dressings. Design: Suzanne Murphy

piece is logically and correctly placed. Only the designer can do this correctly, since it is she who knows what the shots are supposed to be. After the cast and director arrive and the first camera rehearsal is over, the director may come into the studio to ask that certain pieces be moved.

Once agreement has been reached on the final dressed look of the set, the props people will record this for continuity. This may involve sketching the plan of the set and labeling the locations of dressing pieces on it, or it may involve shooting dozens of Polaroid snapshots and putting them in the master props book, or both.

The procedures are essentially the same for location shoots, except that the conditions are a bit more primitive and the need for absolute accuracy and thoroughness on the part of both the design team and the props crew is greater .Often, it is a long drive back to the studio and a forgotten prop can cost a lot of money in lost time.

The preparation for a location shoot may begin several days beforehand, and the design staff, set dresser, and/or prop crew head may confer several times just to make certain that nothing has been overlooked. The props must then be packed to travel, so that there is no risk of arriving on location to find that a critical dressing piece has been broken. A good prop crew will also travel equipped with a tool kit and good assortment of adhesives, spray paints, and tape for emergency repairs. Even so, it is wise to know where the nearest hardware store is to the location.

Again, the best course of action on location is to take continuity shots for inclusion in the props book, even if the production has a continuity person with his own camera there at the same time. Props and dressing continuity is best done by the design and props people who know what to look for and will have to work from the photographs later.

In addition to the obvious things in a props crew's tool kit, there are some special items that can be lifesavers. The most important of these is antiflare spray, which is a dulling spray that will soften the glare from reflective surfaces and will clean off without harming the surface finish. Matte black spray paint can be useful (if you're sure the piece will not be harmed by it), and gaffer's tape stolen from the lighting crew is indispensable (it is flat black and has a special adhesive that peels off without damaging the surface finish). This can be used for glare elimination, for repair of props and even costumes, and for mounting things onto walls without destroying the paint! An assortment of nylon monofilament fishing line is indispensable for hanging things, and a selection of finish nails, which can be driven into walls without leaving visible marks (if you choose the location carefully) and can be bent into an S shape to make picture hangers.

Chief among the resources useful in dressing location sets is a good imagination and a fund of resourcefulness. Everyone has stories of unusual applications to which ordinary household items have been put to bail out a production; you will collect your own favorites.

Makeup and Costumes 16

Roles and Responsibilities

Like their counterparts in lighting, costume and makeup people operate separately from the production design staff, although closely coordinating their efforts.

Typically, in addition to the production designer, there will be a costume designer attached to a given project. Since costuming is a highly specialized craft involving quite different skills in its execution than those required of set construction personnel, the costume responsibilities are managed by a separate group. In an institutional production facility, this may be a department within the organization; in an independent production, there will be a private costume house contracted for services.

Costuming is closely tied to fashion, and is often relegated to the selection (*pulling*) of items of clothing off the rack. This may mean traveling to clothing shops with a presenter's measurements (or with the presenter) in hand and thus assembling a wardrobe to be used in a series of broadcasts. For this reason, the costume staff in a television facility may often be referred to as *wardrobe* staff rather than as costumers. At the high end of the market, such clothing is often donated by the manufacturers in exchange for an on-air mention or a credit in the final *roller*.

Closely allied with the costume department (and often located nearby) are the makeup people. The progress of the talent on shooting day is from makeup to wardrobe to studio, and the efforts of the people in these departments must dovetail to work effectively.

The makeup artists must know what the costume will look like in order to know how much of the body needs makeup and what type and color to apply. They must also know the skin type of the performer to avoid the disastrous consequences of allergic reactions or the need for constant retouching. Someone whose skin is especially sensitive to the chemicals found in some types of makeup may break out in a rash or swell up like a balloon. A person with especially oily skin may actually absorb the makeup under the heat of the lights and need constant fixing.

A person with especially pale skin may need to be made to look more robust or (for a specific role) a person with a dark complexion may need to be made paler. The specifics of these aesthetic choices are often made by the costume designer and coordinated with the overall look of the costume

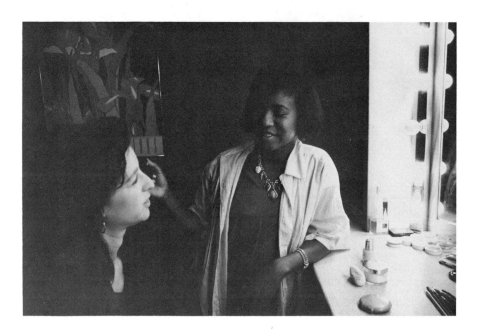

Figure 16.1 A makeup artist at work. New Jersey Network

scheme. This may be derived from historical research or from some other rationale, but a good costume designer will know why the makeup palette should favor a certain set of shades and apply that logic consistently. A piece set in the eighteenth century, for example, would tend to use a pale base color with accents of white and rouge on the cheekbones. Though patently artificial, this look was popular at that time and, together with fake beauty spots, was considered appealing.

The pivotal person in these areas is the costume designer, whose job is often as complex as that of the production designer (and sometimes more so). While costume design requires the same sort of knowledge of history that set design does and the same sort of drawing and rendering skills, it takes a different sort of personality. Settings tend to be larger in scope than costumes and thus, while also requiring a level of detail, do so in a relatively larger way. The detail on a costume tends to be quite fine, extending all the way down to very small patterns of embroidery or lace or to the buttons or stitching. While sets may be shot in relative close-up, costumes are nearly always shot in the same focal plane as the actors' faces, and thus show every tiny detail.

Costume Organization

As with production design, costume design often requires the designer to spend more time organizing and budgeting than being creative. If the wardrobe department is in-house, it reduces the amount of time lost to traveling and encourages a closer working relationship between the costume construction crew and the costume designer, reducing the level of inefficiency. Increasingly (as is also the case with settings), the costume execution task is being jobbed out to independent contractors and this is the situation most designers are likely to find.

The relationship between the costume designer and the production designer should be close and, as with the other creative personnel, they will feed each other's ideas, eventually arriving at a satisfactory compromise solution. The interplay between these two is probably the strongest between creative staff, because both are dealing with color and texture and working within the same time parameters.

In all likelihood, the costume designer will have studied her craft in a theatre school, as will many production designers. As noted, she will spend much of her time doing legwork and managerial work (like her counterparts in production design). She must know how much work is likely to be required for the completion of each item of wardrobe and thus accurately estimate the total cost of the production and the amount of time between inception and completion.

To effect this managerial task, she will have done a thorough analysis of the script requirements and developed a costume plot for the entire production (or shooting period, if it's a series). This can involve a very high level of complexity: A single character may change clothing dozens of times in the course of a few episodes, and each item of clothing may consist of a number of specific pieces (even down to the underwear, if the script calls for it). Articles of costume dressing, such as pocket handkerchiefs, socks, cufflinks, bracelets, watches, and such, must all be found or made, stored together, cataloged, and labeled. Then they can be tried on the actor (during a *fitting*) and altered as necessary.

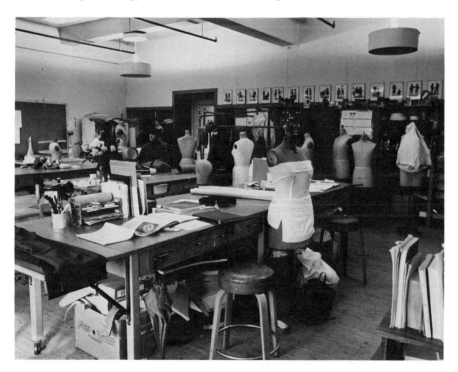

Figure 16.2 A working costume shop.

As the process evolves, the costume staff will collect measurement charts on each of the performers, recording all the pertinent details of body shape and size. Virtually every structural dimension of the body is recorded: Arms are measured for inseam length, outseam length, and circumference at various points; the rest of the body is divided into levels of circumference measurements. When the costume people are done taking their measurements, they have something approximating a carpenter's construction drawing for each performer, allowing them to reconstruct that person theoretically whenever necessary.

From the measurement chart, they may work on an adjustable tailor's dummy, fitting the costumes to the correct dimensions and making the required alterations without needing to call the actor in for fitting until the costumes are nearly finished. The finer adjustments, such as cuff and collar sizes, must necessarily be made with the performer present, as should adjustments for such things as freedom of movement and restriction of blood flow.

The costume designer will have constructed a master plot that indicates the costume to be worn by each performer in each scene. This usually takes the form of a chart (often a very large chart) with the names of performers or characters arranged down the left-hand side, and the scene numbers arranged horizontally across the top. In this way, any actor/scene combination can be easily cross-referenced when it comes time to assemble the costumes required for a given performer on a particular shooting day.

The costume items required for a given scene are pulled off the storage racks and hung up on rolling racks, arranged by performer name and, within that category, in the order required for the day's shoot. Most such racks also have shelves on top for hats and shoes; items of jewelry and other small accessories are put into plastic bags and attached to the appropriate hangers. These rolling racks are then pushed into the studio or loaded onto a truck for shipment to the location or studio, if the costume shop is not nearby.

The costume designer will usually be in contact with the wardrobe supervisor at the initial stage of design presentation, seeking help with the cost and time estimates and help with solutions to anticipated difficulties. Once the execution phase is underway, the designer will participate in the selection and approval of various off-the- rack items and suggest possible alterations and adaptations. Again, during the fitting process and while the costumes are being loaded into the dressing rooms, the designer will be on hand to double-check and help anticipate any problems. During the shoot, she will likely be running between the studio floor (where the effect of a given costume on camera can be observed on one of the floor monitors) and the wardrobe department or dressing rooms, making sure that the best possible on-camera effects are being realized.

Makeup Organization

One of the most important functionaries in television is the makeup artist. You'll remember that the medium inherently distorts what it presents and that the efforts of the design personnel are often directed to nothing more than making a real person *look* real on camera.

Human skin is often one of the least photogenic things you can put in front of a video camera. It is frequently rather oily and highly reflective. Who among us has not seen a bald man's head shining like a cue ball on television? Skin is wrinkled, freckled, spotted, shiny; in a medium such as television, we are often called on to correct these faults and make the faces presented to the camera conform more closely to the popular ideals of attractive appearance, especially in the more idealized formats such as soap operas and variety shows.

Even in the case of the evening news show, a great deal of makeup skill is required. Although the news is meant to be a presentation of reality, we still like our presenters to look better than average people. To some extent, perhaps, the degree to which they look appealing to us is related to their credibility. It is certainly true that the faces shown in shot are a part of the overall visual package and work better when presented to a consistent level of finish with the setting and the graphics that may accompany them.

Like costuming and lighting, makeup is both an art and a science. Apart from knowing skin types and avoiding disasters of allergic reaction, the makeup artist must also (like the production designer) know how the medium will distort the real thing placed in front of it and the degree to which correction is possible and necessary.

A bald man with a dark complexion and oily skin is perhaps the most extreme example. The video camera can accept a *contrast ratio* of approximately 30:1. This means that the brightest part of a shot should not be more than 30 times as bright as the darkest part. Frequently, the intense reflection from the top of the man's head relative to the dark parts of the face below, will exceed that ratio. We can do some correction by placing lighting sources carefully, but the lion's share of this task will fall to the makeup person. By judicious application of a translucent powder, she can reduce the reflective quality of the skin for a more matte appearance and perhaps, by adding a slightly lighter shade of powder, make the overall skin surface slightly less dark. Of course, people with an oily complexion also tend to perspire more easily and the hot studio lights will aggravate this; as a result, the makeup person will often be standing by in the studio to add powder when necessary and to wipe off the accumulation of sweat. This may mean dodging between cameras to get into the set during commercial breaks if the show is live, or between takes if it isn't. Agility is one of the requisite qualities of a professional makeup artist!

In the normal course of most television production (at the local level), makeup is almost exclusively devoted to problem solving like this. At the top end of the industry, however, there is much more scope for creative expression.

Since, to a great degree, television is a closeup medium that relies on subtleties of facial expression to convey much of its message, the face of the performer has become an important palette upon which the makeup artist can do much to reinforce the message. In the context of a dramatic piece, the job of the makeup person will often involve a greater amount of visualization and creative expression than is normal in the other types of programs.

Often, of course, the task is rather mundane, limited to making ordinary people look ordinary, or ordinary people look like somewhat younger ordinary people. There is, however, a branch of makeup known as *prosthetics*, which is substantially more fun. Prosthetic makeup involves various sculptural techniques and materials that can make the face (or other body parts) of an actor look radically different.

Various chemicals can be used to make the skin pinch together to look like a knife scar or a stitched-up wound. Various rubber-based compounds can be built up over the face into virtually any shape; a person can be given a completely different face at will. Working with a mold made from the actor's face, the makeup artists can build a new face from latex rubber, to be stuck onto the actor prior to shooting. This does, of course, take a number of hours to apply and is uncomfortable in the extreme under the hot lights of the studio. The principal use of this technique is to make a younger actor look substantially older, as was done some years ago in the film *Little Big Man*.

The makeup staff are also the people who take care of hair. While the wigs are sometimes acquired, stored, and maintained by the wardrobe staff, they are often fitted and altered by the makeup people. Alongside prosthetic makeup, the use of wigs can make astonishing differences in the appearance of the person. Hair color, thickness, and size can be altered rather easily by skilled wig artisans and with somewhat less discomfort to the actor than the application of a latex face!

As with costumes, the makeup supervisors will prepare charts that cross-reference character names with scene numbers and certain special makeup looks. In the case of a script that involves flashbacks, or covers a large time transition of any sort, the evolution of the actors' facial makeup treatments must progress in lockstep with the changing costumes. Thus it is that the makeup staff will work very closely with the costume staff and the costume designer. Just as is the case with costumes, settings, and props, a certain amount of research is necessary. Often, this will involve finding photographs in old books or magazines, or (if the period is prior to the introduction of photography) in books with color plates of portrait paintings. Depending on the organizational structure of the television facility and the working relationships of the costume designer and makeup people, this type of research may be done by any member of the group. Technically, the costume designer is responsible for the look of the makeup and should only designate responsibility for research and creative input to those he trusts with his reputation.

Summary

In the best of all possible worlds, there is a costume designer assigned to every production. In these days of shrinking budgets, this is less often true. Frequently, the production designer will be expected to design costumes as well as sets and props. In many cases of low-end production, this is not impossible: The average talk show can be costumed off the rack by someone with a good sense of fashion and taste and a grasp of the image to be projected by the program. At this level, the most formidable obstacle often is the disparity between the self-image of the program's presenter and his or her physical reality. Once again, the need for tact raises its familiar head.

However, it is still normal practice to hire a costume designer for programs involving any complicated dramatic script, and the few instances in which you may see no costume design credit are those in which the production designer actually preferred to design the costumes.

In either case, it is incumbent upon the production designer to have at least a working familiarity with the practice of costume design and makeup and, especially, to understand the personality differences connected to the different kind of creativity entailed in these crafts.

17 Lighting

Roles and Responsibilities

You will recall that, from a design perspective, the primary purpose of the talent is to reflect light into the lens of the camera. In the strictest visual sense, that's also true of the set, props, and costumes. Without light, there is no image for the camera to see; if you think this makes the lighting designer an important person, you're quite right.

The craft of television lighting is often regarded as a technical one and, to an extent, this is true. Lighting requires a comfortable familiarity with the physics of electricity and optics and the safe practice of lighting equipment operation. In addition, lighting for television requires a familiarity with the electronic components of the system which interpret and record those optical images that the other creative types work so hard to put in front of the camera.

Television lighting systems involve the use of hundreds of thousands of watts of electrical power. A typical studio lamp will draw 1000 to 5000 watts or more in normal use, and there will be anywhere from ten to more than 100 in use on a given shooting day. The potential for disaster, in terms of serious bodily harm or death, is enormous, and there are many safety features and specific practices that help to ensure against that. A seasoned and careful crew and an experienced designer are essential to the safe use of studio lighting equipment.

From the video perspective, the lighting designer (or lighting director) will need to understand the requirements of the cameras and other electronic image processing equipment in order to get good clean pictures at the output end of the system. There is no substitute for a competent engineering staff, but the lighting director needs to be able to read *color bars* and make adjustments to the monitors (to ensure that what she sees is what she gets) and to read the *waveform monitor* to ensure that the video levels are adequate to produce a clean image without *noise*.

Because of the need for such technical competence, it is sometimes the case that the lighting director comes from a technical background and approaches the job from that point of view without really considering the aesthetic aspects of the job. In such a case, the lighting tends to fulfill only the basic requirement of illumination and often goes over the top in that regard. If you want to see this sort of lighting, watch virtually any prime-time sitcom or any locally produced news show: The lighting tends to be

Figure 17.1 Studio lighting equipment hung from a grid. The news set at KYW-TV in Philadelphia.

very high-key and quite flat. It is unfortunate that this sort of lighting still predominates in much of television today.

The lack of opportunities for people training for the profession to study lighting design for television means that most lighting designers are still being promoted from the ranks of the technicians without any design training at all. The result is that principles of sculptural lighting, the use of color, and the interpretative possibilities of the medium are being overlooked by many lighting people (through no fault of their own). This is of-

Figure 17.2 Color bars and waveform monitor. Can be useful to the lighting designer as well as to the engineers.

Figure 17.3 A green bed at Warner Brothers, Burbank

ten a source of great angst for the production designer because his set design is largely at the mercy of the lighting design. It is of little use to design an intricate setting with lots of delicious texture and sculptural shapes only to have it lit with banks of white floodlights cranked up to peak output. A truly bad lighting job can ruin a design that would otherwise have made Rembrandt weep tears of envy. It can make Mona Lisa's face look more like Betty Boop's despite the best work of a top-notch makeup artist.

Do I make my point? For your own protection, it is wise to know a lot about lighting. Technically, the production designer is responsible for the overall look of a show—even if the lighting ruins it.

While it's not generally feasible to do your own lighting (there are only 24 hours in a day), you may have a good deal of say in the choice of a lighting director and you'd best be able to exercise that intelligently. Alternatively (in an institutional situation, for example, in which the lighting director is *assigned*), you may need to be able to talk intelligently with that lighting designer in order to make tactful and effective suggestions as to the look you want—making it clear that you also know how it can be achieved. Of course, there is always the ideal situation in which the lighting designer is top-notch and actively solicits your ideas and opinions: in this instance, a comfortable knowledge of lighting on your part will ease the process of collaboration and result in mutual satisfaction.

Rudimentary Lighting Principles

The most basic requirement of television lighting is the provision of a sufficient amount of light to stimulate the chips inside the cameras. This is not difficult.

Without getting into the nit-picking details, we need only know that any studio (no matter how sparsely equipped) must permit us to position any of the lighting instruments anywhere we want. Any studio will also have dimming capabilities and will have an assortment of hard and soft light sources (called *spotlights* and *floodlights*, respectively). Spot sources are typically used to light faces and can be made to light only a very small area if necessary. Flood sources give a softer sort of light, are used to soften shadows or to light background surfaces, and cannot be controlled as to coverage areas (they throw light everywhere).

The lighting director usually comes equipped with a *light meter*, which measures the intensity of light in a given place and provides a numerical reading. The light level reading may be in units called *lumens* or it may be in older units called *footcandles*, but either one is an arbitrary unit of measure and is used in the same way by the lighting crew. In the early days of television, the factory specifications that came with the old tube cameras called for a general illumination level of 250 footcandles. This is a lot of light and a lot of heat, and the general acceptance of this as a rule of thumb resulted in some really flat faces and some very uncomfortable studios.

Over the years, two things happened: (1)Lighting directors who were less than happy with the pictures they got began to experiment with lower levels, and (2) the camera manufacturers developed more sensitive tubes (and, later, chips). The result is that now, while the engineers may need as much as 200 footcandles of light on the charts they use to set up the cameras, perfectly good shots can be achieved with substantially lower levels of illumination after the setup has been completed.

Key and Fill

At its most basic, television lighting seeks to make the face in front of the camera look real and immediate. This may involve a number of specific goals: the softening of hard lines or hard shadows, the filling of wrinkle lines to hide the effects of age or weather, or *kicking* the eyes to add sparkle. These are the more subtle of the basic tricks, though; the real, and more difficult trick is to defy the reality of the cathode tube and make a two-dimensional representation have the feel of three dimensions.

The technique used for this purpose is *key and fill*. The system used is built around two light sources hitting the face from opposite sides and at different levels of intensity. They are not usually *directly* opposite each other; rather, they are placed approximately 45 degrees to either side of an imaginary centerline drawn through the head of the talent. They are also placed approximately 30 to 45 degrees vertically above the plane of the eyes. These positions will usually give a good sculptural effect on the average face; all that remains is to identify one source as the *key* (set at a higher level of output) to define the shadows and the other as the *fill* (set at a lower level) to soften the shadows and provide information about the parts of the face that are in shadow.

There are many adjustments that can and must be made depending on the architecture of the particular face being lit. The basic idea behind

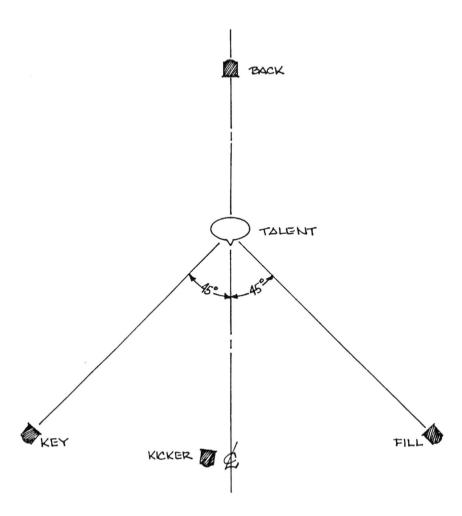

Figure 17.4 A typical key-fill-back set-up

the system, however, has proven itself to work over the years, with the addition of some other elements.

Backlight

The most important addition to key-and-fill sources is the backlight. This is typically placed directly behind the head, and at a fairly steep angle (high up) in order to avoid throwing light directly into the camera lens. The backlight outlines the head and shoulders and, in so doing, tends to define the shape of the person and appear to bring it forward, away from the background. Just as the key-and-fill sources accentuate the shape of the face and enhance the appearance of three-dimensionality in the plane of the face, the backlight accentuates the separation between the head and the set behind it and enhances the effect of depth at that level.

This combination of key and fill and backlighting is so easy to use and so effective that it is the standard formula for virtually any talking head setup in any part of the world. With a little care and thoughtful plan-

Figure 17.5 Key and fill and kicker in use at KYW-TV

ning, the angles and levels of the different sources can be adjusted to simulate fairly natural lighting conditions and still achieve sufficient overall levels to get good pictures and give the camera (and the viewer) enough information about the face in shot.

The key source is nearly always a *Fresnel* spotlight and the fill may be either a Fresnel or a soft source. The backlight is also a Fresnel (for control purposes). Another Fresnel (or other hard source) may be set at a low angle from the front and used as a *kicker* to highlight the eyes. A soft source may also be placed at a low angle in front of the face to soften the harsh lines of age or a particularly angular facial structure.

Dressing Light

You may recall that the term *dressing* was used to refer to placement of properties and set pieces on the studio set to simulate a realistic level of detail. The light that is used to illuminate the setting is also called *dressing*, and for the same reasons.

As I noted earlier in this chapter, the extent to which the texture, color, and detail of the set are visible is largely due to the effectiveness of the lighting. Simply lining up a row of intense soft light sources and calling that set lighting is far from adequate—even though it is still frequently done. Although facial lighting does not, as a rule, use color filters, it is often helpful to add color to the lighting on the set to enhance the effect of the lighting. It is also very important to adjust the angles of the set dressing light sources to enhance the shapes and textures designed into the background pieces. The combination of color and angle, when carefully and correctly used, can correct for the difference in color response between the eye and the camera and can intensify the effect of depth, form, and texture.

Figure 17.6 Dressing light. Note how the walls of the set are carefully highlighted and how the canister lighting is built into the headers for effect. Design: Frank Schneider

Just as the key-and-fill system offers substantial opportunities to enhance the sculptural effect on the face, so the same principles may be applied to any three-dimensional form. A hard source set at an angle to a surface will accentuate the shadows and provide maximum definition of form. If this is too harsh, softer sources can be added in from another angle to lighten up the shadows and ease the overall contrast. If the effect you want is that of a flat surface (say, the way the sky appears to the eye), then there is nothing better for it than to use a bank of soft sources straight in from the front. If the set has texture and you want to enhance it, use hard sources at steep angles from high overhead. If you want your set to glow, that's easy: just build it from some sort of translucent plastic or thin cloth and light it from behind.

You will remember that the tubes or chips inside a television camera see colors a bit differently from the human eye. These differences are rather subtle, but as you work with TV equipment you will get a feeling for the differences. One of these is a tendency on the part of cameras to react strongly to blue; this can be very useful when going for a sky effect and downright maddening when you've painted a wall a subtle shade of blue and it screams at you in electric blue on the monitors. Problems such as these can be corrected by washing the offending surface with the *complementary* (opposite) color in light, thus neutralizing some of the undesirable effect. (This is, of course, only useful in changing the shade of a color in a fairly limited way.)

The use of color, angle, and intensity variations to achieve specific effects must be understood in more than a general way by production designers as well as by lighting designers. As a production designer, you are responsible for the look of your sets and you must know what effects are possible and how you can achieve them before you commit yourself and the production company to achieving those effects. For this, you need to know something about the mechanics of studio and location lighting as well as the aesthetics.

Lighting Mechanics

A table lamp has two basic parameters—it is either on or off. You can move it from one area of the room to another, of course, and some may have three-way bulbs or dimmers to adjust the brightness, but the average table lamp is a very unsubtle instrument. A studio lighting instrument is very different; it is much more purpose-oriented and technically sophisticated.

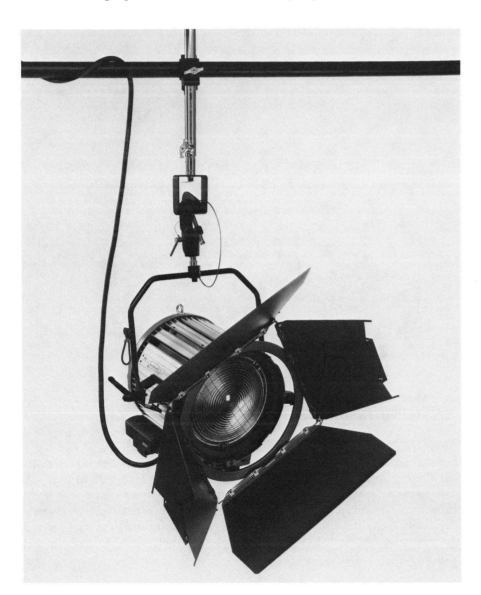

Figure 17.7 A studio fresnel. Note the hanger and the barn doors. Photo courtesy of the Arriflex Corporation

Figure 17.8 Film-style lighting equipment. Floor-mounted instruments, C-stands, and flags are used instead of a grid, gels, and barn doors. Photo courtesy of the Arriflex Corporation

Most studio lighting systems have dimming capabilities. Typically, this is built in and *hard wired* into the lighting distribution system (which delivers power to the outlets on the lighting grid). When the light source is plugged into a *dimmer circuit* and the lighting board operator slides the *dimmer control* lever up, the light comes on to a certain percentage of its full output (depending on the position of the control lever). Thus, it's possible to vary the intensity of a light source infinitely from 0 to 100%. If the dimming system is computer controlled, the operator need not necessarily physically move the control; she may simply program that instruction into the control computer so that it will issue the instruction to the dimmer unit directly. Whichever way, the effect is the same.

On location, dimmers may be too much trouble to carry along; there is another method for dimming a light source. This method was borrowed from the film industry and involves the use of special *gels* (plastic filters) placed over the source. These are called *neutral density* filters because they cut out a specific amount of light without affecting its color in any way. They're available in a variety of degrees of darkness and are much less trouble to carry out to locations than are portable dimmers.

Television lighting equipment is also capable of being *aimed* and *focused*. Aiming and focusing are usually accomplished at the same time and the routine is fairly standard. Each lighting instrument is hung from a pipe *grid* or from an adjustable *hanger* unit and is connected to the hanging system by means of a *yoke*. This yoke permits the instrument to be swiveled horizontally (*panned*) or vertically (*tilted*). The electrician points the lens (or front) of the unit at the person or thing to be lit and locks it into position. He then focuses the unit, adjusting the optical elements (via external controls) to make the beam of light narrower or broader, as necessary. Each type of light source has different optical characteristics and different mechanical means of adjusting focus. The soft sources tend to be more difficult (or even impossible) to focus, but they also tend to be used to light extreme background areas where the tightly controlled beam of light is not needed, such as *cycloramas* and backdrops.

If the beam of light needs to be squared off rather than rounded, that's possible too. The Fresnel spotlight and most of the *open-face spotlights* (without lenses) can be fitted with steel shutters called *barn doors*, which

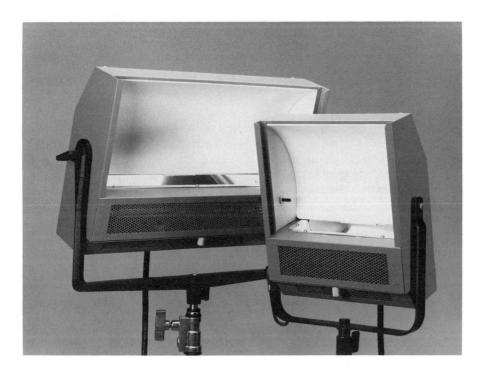

Figure 17.9 Soft lighting sources. Photo courtesy of Arriflex Corporation

Figure 17.10 An ellipsoidal spotlight (Leko)

permit the pool of light to be shaped into a rectangle by pivoting the barn doors (which are mounted on hinges) in or out. Another type of lighting instrument, which is optically more advanced than the Fresnel and designed more for theatrical than television use, is the *Leko*. It has internal shutters (which slide rather than pivot) to make the beam of light square or rectangular (or a trapezoid or a parallelogram). The Leko gives a very hard-edged beam of light and is typically used in television only where the light makes a sharp pattern on a surface or has to be very tightly controlled. It can also be used as a pattern projector, when a metal cutout (*gobo* or *cucaloris*) is inserted where the shutters would be (called the *gate*).

A myriad of lensless instruments is available in varying degrees of hardness. Technically, only Lekos and Fresnels are truly hard sources; anything with no lens may rightly be regarded as a soft source. Still (especially on location), some lensless instruments are used as key lights and thus really act as hard sources. These are small, light, and very powerful and are typified by the Ianiro portable units (although several other manufacturers make equally fine portable instruments). They are, to a limited degree, focusable, and can be mounted to virtually anything by means of a *gaffer grip* (a special clamp). On location, these are typically used open for hard sources and gelled with a *diffusion* material to make them softer. A carrying case with a number of these instruments, a selection of portable stands and gaffer grips and an assortment of diffusion and neutral density gels (along with miles of cable and adapters and such) will work lighting magic on an otherwise difficult location.

When we work in the studio, however, the distances and surface areas tend to be greater, and more powerful equipment is required to light the set properly. The soft sources used in the studio (like the hard sources) tend to be heavier and bulkier and much more powerful than their location counterparts.

The most basic of these, and the oldest in design, is the *scoop*. This is simply a concave circular reflector with a lamp in front of it and a slot on the front edge to hold a gel frame and gel. It can be aimed but not focused and may be used for very soft front fill light or for lighting large background areas, such as cycloramas or backdrops.

Also used for these purposes are a number of units generally referred to as soft lights. These take a number of forms, but the unifying principle is that direct light from the lamp inside is never visible from in front of the instrument: All the light that comes out is reflected off a slightly diffused silver surface first, ensuring that it is quite soft. These can be used in place of scoops, but are really not very good for lighting backdrops.

The ideal instrument for large vertical surfaces is either called a strip light or a cyc light. Functionally, these are similar; a strip light tends to be rather long and narrow and is only good for lighting a long flat surface. The cyc lights tend to be shorter units and can be easily arranged in a curve for lighting a cyclorama that is curved around the corner of the studio. They are also, of course, capable of lighting flat surfaces equally well. These units

Figure 17.11 Lensless instruments in a portable light kit. Photo courtesy of Arriflex Corporation

are typically permanently gelled and wired in sequential groups of three or four lamps: This permits the *primary colors* (red, green, and cyan) and white to be mixed in varying proportions to make whatever color is desired on the backcloth.

This is a very quick tour of lighting mechanics. If you are beginning to think that television lighting is technically complex and are beginning to develop a healthy respect for those who are good at it and have mastered both the technical and artistic parameters, that's good and will serve you well!

Lighting Aesthetics

As daunting as the technical side of lighting may seem, mastery of the nuances of its artistic application is somewhat more difficult, as evidenced by the smaller number of people who have done so. Unlike painting or sculpture, designing light is an extremely esoteric craft and requires a very active and vivid imagination.

The tough thing about light is that it isn't really there: If you build a flat, it stays built and if you paint it blue it stays blue. If, However, if you aim a lighting instrument into a certain space and gel it blue, no one knows that until someone or something moves into the light. Light is only evident when it reflects from something to the eye and it takes on the form of the thing it hits, rather than having an apparent form of its own.

Light does have form, of course. Once the light source is fixed in place and the shutters or barn doors adjusted, the beam of light being emitted is given a shape that will not vary unless the source is moved. The difficulty is that only the lighting designer and crew know what that shape is and until it hits something that information remains a theoretical con-

Figure 17.12 Cyclorama lighting. Note the box-like cyc lights with four lamp compartments, each gelled a different color.

Figure 17.13 A lit face. Motivated lighting attempts to create the effect of real light sources.

struct in *their* minds. As you can see, documentation is extremely important to the lighting designer; while the design itself is the interaction between light and form in the studio, the light plot is absolutely critical to the understanding and interpretation of the system of light that constitutes the design idea.

The experienced eye will be able to look at a light plot and discern patterns of instrument placement that will reveal the mechanics of the design—the way in which the different light sources are intended to interact with each other. Practice at lighting will help the designer develop the ability to recognize typical angle relationships that occur frequently and enable the reader of a plot to divine the purposes of each of the instruments in relation to the others. An understanding of the key and fill and backlight system will lead you to look for groups of three sources arranged in that angular pattern. Spotting backing flats or cycloramas will cue you to look for the kinds of soft lights that are typically used to wash these surfaces (or, alternatively, their absence—which would indicate a different creative choice and prompt you to ask why).

The sculptural tools of light are angle and color. These are also the emotive tools, and the two functions are so closely interlinked when making a statement with your lighting that it is often impossible to make a dis-

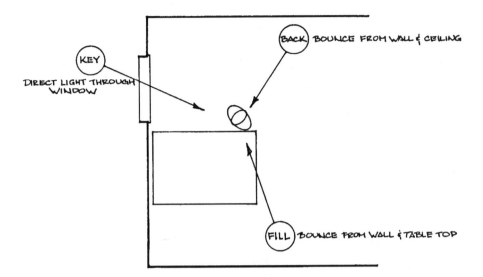

Figure 17.14 A plot for the lit face

Figure 17.15 A wide shot of the lit face. Note the contextual elements that motivate your choice of lighting sources, angles, and colors.

tinction. Almost any visual image is evocative of a mood; the amber glow of the sun striking someone's face through a window at a low angle tells us that this is probably a summer evening and triggers in our recollections a catalog of associations that cue a mood reaction. Physically, this isn't a complicated phenomenon: A source of light of a given color strikes a form at a given angle and reacts both with the form and with its own colors to create an image in the observer's eye. Pretty dry stuff, if you take out the emotional associations, isn't it?

In lighting design, however, the emotional associations are the elements that make it an art rather than a science and make the design a design rather than a formula. If you want to illuminate a face, apply the three-point formula exactly as described; if you want to design a light plot, start with the three-point system and build from it to something better.

Much of TV lighting is realistically derived. Often, the task will be to recreate a mood from a specific lighting situation, such as the sunset scene just described. The key to this sort of design is to begin with a physical analysis of the situation and to build on that.

The key-and-fill system always works when the sun is involved; since the earth has only one sun, that naturally becomes the key source. Because it's late evening, and the room has a window, we may naturally decide to place the key source at a low angle outside the window and gel it amber. (You may note that, in different seasons and at different times of day, the light coming from the sky is a different color.) If there are curtains in the window (lace, perhaps) or the sky is hazy, the sunlight may be a bit diffused and we add a diffusion filter to the sun source.

We then proceed to find positions for the fill and back sources. What are the fill sources in a real room? The walls, of course. What color are the walls in this room? What color would they reflect if struck by amber sunlight? What angles are they to the face we're lighting? Where is the closest source to the face? (This will likely be the fill source, won't it?) From where will the backlight come? If there's no obvious source for a backlight, can we do without it? A lot of questions need to be asked to do a lighting analysis, but they're all really asking the same question: What are the natural sources of light in this situation?

As with set and costume design, the skill required to answer them is a capacity for accurate observation. To be an effective lighting designer, you need to be able to see light in real situations and to remember what it does. You need to develop in your head a catalog of lighting looks and the physical parameters which define them. As your catalog grows, it becomes easier to access specific solutions, and you find yourself spending more time with the subtleties and details and less time with fundamental analyses. Once again, there is no substitute for experience and no better teacher than practice.

The toughest sort of lighting is that which seeks to evoke a mood response without a basis in reality on which to hang it. Dance, music, and variety lighting are often the most difficult for this reason. It is never really sufficient just to throw up a bunch of pretty colors and hope for the best; music is evocative of specific emotions and needs equally specific support from the lighting design to achieve the right effect.

Summary

As with costume design, the production designer and the lighting designer need to be working toward the same goal, and (as nearly as possible) need to be in lockstep with each other. In reality, the entire look of the production is at the mercy of the lighting design and all of the other creative personnel are usually aware of this. The result is that many lighting directors become rather reclusive; this is a natural result of being bombarded with unreasonable (and frequently conflicting) requests from all quarters. If the production designer understands lighting sufficiently to know what is reasonable and what is not, she may prove herself a comfort, rather than an annoyance, to the lighting crew and, may find that they're rather easy to work with. Too many in the studio regard lighting as some sort of magic; they really think that the lighting director can make a 70-year-old actress look 35 or make a green costume red. It's important that the production designer not do this; as the person responsible for the appearance of the show, she needs to make the decision that says, "Light the actress consistently with the scene, and don't worry about her vanity," and have the conviction to stand by it and the expertise to defend herself.

Graphics 18

Parameters

By definition, the area defined as graphics may include anything from simple captions over pictures to the most elaborate artwork.

In most applications, television graphics are supplementary to the more important primary information carried in the picture itself. Probably the most frequent use of graphics is the simple superimposition of a few words over a picture to provide identifying information. This function is called *captioning* and has been a staple of TV news since it became possible to superimpose one source over another. This was originally done by means of a *luminance key*, which relies upon the fact that the system will accept the brighter of two images. In order to key a caption over a picture, it was necessary to apply peak white lettering onto a black surface: When mixed with the picture, this would show white lettering over the picture and picture information over the black background behind the lettering.

The system worked rather well and was used until recently in many production facilities. To get a clean image, large mechanical letterpresses were used to bond white material to black cardboard wherever letters were needed. The card would then be placed on an easel in the studio, lit, and shot by one of the floor cameras. In the earlier years, only a white caption was possible, but later it became possible to insert color into the letters electronically, once the key effect had been achieved.

The system had its drawbacks, of course. The amount of black card required to service this practice was immense; I recall seeing graphic assistants wheeling handtrucks of the stuff into the caption workroom every week and trash bins outside studios filled with stacks of discarded captions. Often, it would take several attempts to get a clean print on the card and many failures would also be thrown away. The work involved was drudgery of the worst sort and the assignment to this task was reserved for the lowest echelons of the graphic department.

Fortunately, we have entered the electronic age and this practice has gone the way of the dinosaur. Computer-driven captioning equipment is the order of the day now and such keys (or *supers*) are all generated electronically and inserted as first-generation video information without the use of a camera. As computer-generated graphic systems were developed for the graphic arts industry, the number of fonts available increased, the resolution improved, and an enormous variety of colors came into use—all generated quickly and simply by a single *cap-gen* operator.

It is currently normal practice for a program to develop a graphic identity, a look that is unique to that program and will identify it easily for the viewer. This will extend to the consistent use of a specific type font, a color palette, and any other graphic treatments surrounding the information. The ABC red-line graphic identifies the output of ABC stations by the insertion of a thin red line below the lettering; such simple treatments have given way to more sophisticated ones, such as graduated washes of color behind the letters, boxes around them, or small graphic symbols adjacent to them. Sports programs, in particular, make frequent use of special graphics to enhance the massive amounts of statistical information regularly inserted into the picture. This has reached an extreme in the television coverage of New York Yankees baseball, in which game stats are displayed in a graphic panel that looks like the architecture of Yankee Stadium. With the microprocessor at the heart of the captioning system, such frameworks for simple information graphics can be designed and memorized to be called up at will and used in conjunction with any words or numbers that may be appropriate.

You may be thinking that the distinctions between captions and serious graphics may be a bit fuzzier than they once were. This is true; once upon a time, graphic artists and caption-generation personnel were separate beings and often members of different unions. The development of sophisticated in-studio captioning equipment such as we now take for granted was cause for some disruption in many TV facilities when contract-negotiation time came around, and such contention was often the reason that some facilities didn't implement the new technology until it was rather old technology.

Personnel

Typically, a major program would have its own graphic designer and a number of graphic artists assigned to it. The graphic designer works with the production designer and the director devising the graphic identity for the program and supervising the preparation of graphic requirements for each installment.

The graphic designer is responsible (along with the production designer) for maintaining a consistent graphic quality and so needs to be in constant communication with (and control of) the graphic artists and anyone preparing caption materials. For a program with complicated graphic requirements, the graphic designer's job may well become one of resource management, much like the production designer's. These days, some programs use very complex animated graphic treatments, which tend to require that the graphic designer spend most of his time in preproduction work supervising the preparation of complicated artwork or animation.

Artwork

The elaborate treatment of simple caption lettering leads us up a step to the more complex world of graphic art. This work is never done in the studio on the day of the shoot; typically, days or weeks of work have gone into the development of the right graphic look and the results will have been the product of concentrated efforts by the director, the graphic and production designers, and sometimes the program producer.

It is not uncommon for the workaday world of news graphics to come up with specially prepared pictorial treatments to go in the over-shoulder box along with a caption of some sort (such as "Fire!" or "Murder!") to identify the content of the story being reported at the moment. This is especially true on a breaking story for which there is no live footage of the event available. Often, too, different kinds of charts and graphs are used to illustrate numerical concepts, financial trends, results of polls, and so on. All of this means that there are graphic artists at work in the news department, right up to the moment of transmission, producing high-quality graphic support images on very short notice.

This sort of sophisticated work can only be done on such quick turn-around because of the advanced technology available to the graphic staff today. The paintbox hardware typically at the disposal of the television graphic artist enables her to quickly do subtle shadings, line drawing, lettering, and pastiche techniques to a very high standard of finish. Since these are not done in hard media on pieces of cardboard, time is no longer wasted setting them up in front of a camera and they can be put on the air within seconds of being completed, if necessary.

There are a number of different ways of encoding picture signals, and computer-driven graphic systems often use different encoding from that commonly used in video studios. However, there are, conversion devices that enable the output from a paintbox system to feed into the studio system and that can be brought on-line immediately. Alternatively, such information can be recorded on disc and played back into the system from the stu-

Figure 18.1 A high-definition paintbox graphic. High-resolution graphic work being done in the 9 x 16 HDTV format. Note the function boxes superimposed at the bottom. Courtesy of Quantel Corporation

dio control room. Either method is infinitely more flexible and responsive and less wasteful of time and resources than the old hard-graphic one.

As discussed earlier, the new graphic equipment is also capable of producing three-dimensional and animated effects much more efficiently than was possible previously. There are many generic flip-flop, cube, pixelization, and other motion effects that are supplied with the hardware, preprogrammed. In addition, and at somewhat greater cost in terms of time and effort, virtually any desired animation effect can be achieved with most of the high-end systems now available to even the smaller-market television stations. Many of these effects are showing up in weather-report graphics, illustrating the progress of storm fronts or hurricanes across maps or in cheerful graphics accompanying the relevant statistics for long-range forecasts.

The rapid development of software packages for such graphics applications has opened up exciting prospects for the people working in this field. The hardware and software together make a magnificent tool for the graphic artist with a quick mind and a strong imagination, and that raises my next point!

Aesthetics

As is true of other design positions, the most valuable asset any graphic artist or designer can possess is a creative imagination. While it's difficult to design a studio set without exhibiting some creative ability (no matter how small), that does not necessarily hold true for some graphics situations.

There is a great deal of sophisticated graphic software available now and many a blindingly impressive effect can easily be called up from the program menu, pre-programmed into the system. This is not necessarily a bad thing; when it's the eleventh hour and something must be put on screen, the ability to call up a preprogrammed effect can be a lifesaver. The ability to program into one's graphic system a catalog of effects is a considerable part of the benefit of having such a system and, as such, is a powerful tool for the artist and designer.

Still, the temptation to mask lack of imagination or initiative by calling up someone else's ideas instead of inventing your own is a strong one. As the software for television graphics gets more sophisticated and the catalog of built-in fonts and graphic effects grows, the greater the temptation for the artist to become an operator, instead. This is often quite unfortunate; the person with little or no imagination can claim to be a graphic designer and sometimes get away with it and the talented designer can take the easy way out even when constraints of time or budget don't require it.

When television is really good, it is always hand in hand with a certain attitude on the part of the creative personnel. This is easy to be glib about, but the best people in this industry are those who think that each program (indeed, every episode of every program) needs to be approached as a unique entity. The object in acquiring such an approach to your work is to think through every project with its own specific identity in mind. To be among the best in the field, you should go back to the window exercise of Josef Svoboda; every detail of the design (set, lighting, costumes, props, and graphics) should, as much as possible, be specific to that show and that moment of that show if that level of attention to detail is warranted.

Thus the graphic arts systems in use in television—just like the scenery shops, lighting equipment, and other support systems—must be seen as means to the end. They're tools in the hands of the creative staff, and it is the creativity those people bring to the use of these systems that makes the whole thing fly.

19

Where Do We Go from Here?

Interactivity

By now you are, I hope, aware of the basic system by which interaction between television and its audience occurs. This is, at best, a crude and cumbersome system born in the late 1940s and, as such is a far less sophisticated system than modern technology is capable of effecting.

Speculating about the coming of HDTV is popular in the industry today. Bear in mind that what we're used to seeing today was described in its infancy as "high definition television" and that the term is relative rather than absolute. The people who are actually doing production in the NHK/Sony 1125 format (HDTV) have begun to refer not to high definition, but rather to high density. This is an outgrowth of the computer industry rather than the broadcast television industry and its purpose was industrial rather than commercial.

In the days of computers that filled entire rooms (or even entire buildings), the computer was conceived as a gigantic problem-solving and information storage tool. As with most innovations, the originators were hardly able to predict the real impact of their creation. The computer as we know it today. The computer has become a tool of communication and it continues to move in that direction. Networking is an accepted fact and is now easily available to the home computer user not just to the business sector.

It is no mistake that the linkup of many computers is known as a *network*: In essence, a computer network is no different from a television network except that it doesn't use public airwaves and is, therefore, not government regulated. Such a network may use ground lines (telephone lines) or satellite links, or (most likely) both.

Without consumer demand, created by the proliferation of personal computers and their evolution into a means of communication, there would not have been sufficient demand and commercial stimulus for the creation of these networks. Without such networks, the potential for interactive television would not be as far advanced as it is now.

It is true that the cable television industry provided a viable alternative to broadcast TV and a means of interaction that is somewhat more

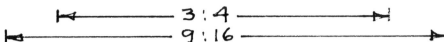

Figure 19.1 HDTV versus conventional formats. Note how much information is lost in the old 3 x 4 aspect ratio. Although it looks now as though the current analog 1125 HDTV format will give way to a digital system of even higher resolution, the 9 x 16 ratio will likely remain

sensitive than the outmoded and unwieldy ratings system. Still, the response mechanisms, when they exist, are not very good when compared with the immediate and thoroughly conversational capabilities of computer networks.

Let's look at this more closely and we'll see how the technology facilitates true viewer interactivity and why it will appeal to the driving forces of the TV industry. A public relations representative for Rebo Studios in New York (one of the early HDTV facilities to adopt the NHK/Sony hardware), posed an example. Suppose you're watching a prime-time serial drama, such as "LA Law." In a set depicting a character's living room, you see a chair that you think would look fantastic in your own house. Stepping over to your computer keyboard, you freeze the picture and use the cursor to identify the chair in question. Next, you ask for the name of the manufacturer. When it comes up on your screen, you call up their catalog and find that the chair costs a certain amount of money, is available in a certain variety of fabrics, and is sold through a certain retail store in your neighborhood. Being a decisive person, you select the fabric, place an order, and charge it to your credit card. The computer verifies the store's receipt of your order and, perhaps, responds with a delivery date.

This is the difference between high definition and high density: High density television carries a picture of greatly enhanced resolution (definition) and, perhaps, of different aspect ratio than conventional TV, however, it also carries much more information (density) on the sidebands, the parts outside the picture information.

Outside the Mainstream: Nonbroadcast Television

One of the predecessors to nonbroadcast television was the corporate teleconference. This has existed at several levels of sophistication and speed of response. In its most basic form, this type of television may be no more than a closed-audience news program. Many companies do this; once a week or so, they will broadcast (via satellite) from the company headquarters to their own facilities in other locations, bringing the employees up to date on new product lines, sales information, and so on. Sometimes, such a one-way broadcast will be presented in an interactive environment—perhaps in an assembly of employees—led by a number of company executives who are informed on the subjects being presented and can take and answer questions. The next step up will be the addition of an audio return, a sound-only link from the remote site to the point of origination. Typically, this will be effected by means of conventional telephone service. In this system, a video signal goes out from the company base to its branch locations and a sound-only signal comes back, so the employees in the branch sites can see and hear the speakers and can ask them questions, but cannot themselves be seen.

The most elaborate and most expensive system is also the most truly interactive. In the top-of-the-line setup, there are video links going both ways, from the base site to the branch sites and from the branches back to the base. This is expensive in the extreme and only rarely used (and even then only by the larger and wealthier companies). Rental fees vary greatly, but the purchase price of a downlink facility (for reception of video signals from a satellite source) costs about one tenth as much as an uplink facility (to send signals up to a satellite). Add to this the expense of renting time on a satellite, the various technical problems associated with satellite linkups, and you have a rather costly and cumbersome system. Still, it costs far less than flying several thousand corporate employees to some distant city for the sort of annual conference often staged by large companies in the past.

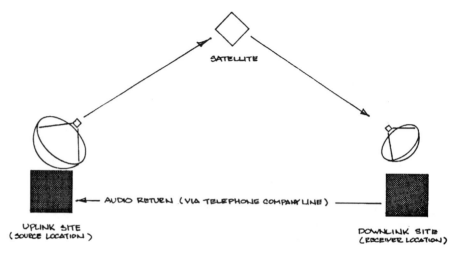

Figure 19.2 Corporate satellite TV. Due to the rapid expansion of available satellite channels, it is now possible to set up a temporary television system on rented links. The return line may be audio only or both audio and video.

Further Technical Advances

Let's add another element into this mixture. On the near horizon, and generating some controversy already, is a video telephone system. In the strictest sense, this is truly interactive video, since the concept of interactivity is really a matter of the degree to which a system approximates a face-to-face conversation. Surely, though, this would never prove a substitute for broadcast television in the entertainment market. Some technicians will say that the picture quality of such a small camera could never approach that necessary to achieve broadcast standards. If you believe that, look at TV coverage of auto racing and ask yourself just how large that on-board camera is that sits over the driver's shoulder. (It's tiny.) If you think it would be too expensive, remember that the one thing that brings the price of sophisticated electronic products down is consumer demand and quantity production. If you think that video telephones will never happen, you're probably wrong.

Where the Designer Fits In

All of this is, no doubt, exciting, but what has it got to do with design? Apart from stimulating the imagination (a process that anyone who aspires to a creative profession should welcome), these things will likely be among the forces that shape the evolution of the medium in years to come. Since the advent of cable television, it has been written on the wall that the centralized production and dissemination of TV product from a single source (Los Angeles) was destined to change. Those who thought of the medium as a static industry have been proved wrong. Those who thought that they could control the evolution of the medium have also been proved wrong. The nature of television is defined by its relationship with its audience and we need to remember that the demands of the audience (although often slow to make themselves known) do drive the system. As such, it is a dynamic thing and one that will continue to defy the pundits who would predict its outcomes and the industry magnates who would control them.

How, then, might the trends in development of the medium affect the designers and other creatives employed by the industry? If anything, the most immediate result of the forseeable changes (based on recent history) will be the continued diversification of both sources and audiences. From three networks and an occasional independent or PBS channel, we now have upwards of 30 channels accessible to most viewers, and there is no reason that this trend will not continue until it reaches its saturation point. This has happened only because viewers wanted it and, if those viewers want it, a more interactive system (possibly based on one of the systems mentioned) may well develop.

In any case, the power and influence of the television monopolies (either the Los Angeles–based industry or a state-supported monopoly such as the BBC) has been diluted substantially. Neither is in a position now to dictate the terms of business, and that has made the market more competitive. This has raised the heat in the kitchen, so to speak; the networks that were once the only game in town—competing among themselves, but in a very limited way—are now required to compete more vigorously just to retain their audience share and the facilities in which the greater share of TV product was created no longer have a corner on that market. This means that the quality of the product has been forced to improve, at least in visual terms, and that the production of television programming has begun to spread out, returning to New York and opening new facilities in Vancouver and Florida. The importance of the creative contribution to programming is being recognized to a greater degree by the industry (since it relates to the competitive edge in a direct way), and support for such services (in terms of budgets for creative departments) is being increased.

While there was a quaint naiveté about the flat, obviously painted scenery of shows like "The Honeymooners," and the awful flat lighting that accompanied them, it is a better situation all around when talented designers can do realistic and accurate settings, lighting designers can take the time to do their work carefully and artistically, and television programs can do believable special effects without breaking the producer's budget. These improvements are due, in large measure, to the fact that the television viewers have become more visually sophisticated and forced the system to upgrade the visual quality of its programming.

Having just attacked the industry pundits, I won't indulge now in any predictions about the prospects for the future. I hope, however, that you, as a designer, will not forget that the audience drives the system and that, in an increasingly competitive market, those who have their fingers on the pulse of the audience will be better able to accommodate changing circumstances and, therefore, better able to survive. Nobody enters the television profession in search of a sedentary life-style; it's an exciting medium and can be quite stimulating if you approach it with the right attitude. I would invite you, instead of me, to do your own postulating about the new developments in technology and audience-industry relationship that will shape the television of the future.

Glossary

Artifact An undesirable aberration in picture quality, such as a color shift, halo, outline around a person in shot.

Aspect ratio The relative height-to-width dimensions of any picture format, expressed as a ratio. (TV is 3 : 4).

Backflap hinge A simple steel hinge used in scenery construction, for hanging doors or (loose-pin) for joining.

Backlight The lighting instrument placed behind a person to outline the head and shoulders and separate the figure from the background.

Balustrade A run of balusters with a handrail on top—such as is found on a balcony or stairway.

Barn door The device mounted on the front of a Fresnel to cut off the sides, top, and bottom of the light beam.

Baseboard (skirting board) The decorative molding between wall and floor.

Binder The glue in paint that sticks the pigment to the surface to be painted.

Brace A telescoping pole that is attached to the back of a flat and to the studio floor to support a flat or run of flats.

Cap gen A caption (or character) generator—a device that builds titles for insertion over pictures on-screen.

Casement In scenic use, a type of window that is hinged vertically to swing open like a door.

Casing The molding that surrounds a door or window opening.

Cel A sheet of clear plastic that has been treated to accept artist's media, such as ink and paint. It is used in film animation.

Centerline An imaginary line through the middle of something, most often a stage set or a symmetrical set piece.

Chair rail A molding around the walls of a room, put there to prevent chair backs from scratching the wall finish, but now largely decorative.

Chip A charge-coupled device; the silicone chip inside a TV camera that makes electronic imaging possible.

Chrominance The technical word for the property of color as distinct from brightness; the word for color itself.

Color bars The field of colored stripes, electronically generated, used to set up a video system for uniform color response.
Color wheel The theoretical wheel composed of wedges of color. *Complementary colors* are those that appear opposite each other on this wheel.
Complementary color A color on a color wheel opposite to another color. These colors tend to neutralize each other when mixed (i.e., red and green).
Contrast ratio The ratio of lightest picture area to darkest: film is 100 : 1, video is 30 : 1.
Cookie See *Cucaloris*.
Cornice The molding between wall and ceiling.
CSO A color separation overlay—the chroma-key effect in which one image is substituted for a specific color in another shot.
Cucaloris (also gobo) A metal pattern used in a Leko to project a pattern in light onto a set or floor.
Cyc light A special lighting instrument designed to light the large flat expanses and curves of a cyclorama.
Cyclorama A large smooth expanse of canvas used to indicate sky or infinite distance; a true cyc is curved and wraps around the foreground pieces.

Degradation The tendency for picture quality to deteriorate with each successive re-recording.
Dependent scenery Scenic pieces that cannot stand on their own and must be attached to another unit.
Detail An enlarged-scale drawing to show details of construction or particularly intricate finishes in set construction.
Diffusion A type of lighting gel that softens the light rather than altering its color.
Dimensioned drawing A construction drawing with all necessary dimensions noted in correct standard form.
Dimmer An electronic device that regulates the amount of power flowing to a lighting instrument.
Dimmer circuit The electrical circuit that delivers power to the lighting instrument in studio.
Dimmer control The control handle on older lighting control boards that is physically moved by the lighting operator to raise or lower the intensity of the light.
Dip to black The practice of fading the picture to full black briefly, then fading up again to indicate passage of time or change of locale.
Downstream A component in an electronic system that is placed after another component in relation to the direction of flow of the electrons is said to be downstream.
Dressing Finishing touches in set properties or lighting that make the whole picture look just right. Such props may be called set dressing and lights dressing lights.

Elevation A scale drawing showing a straight-on view without any perspective line convergence, so that measurements may be taken from the elevation.
Eyeline Either an imaginary line drawn through the center points of the eyes or a reference to the direction in which a person in shot is looking.

FEV A painting technique using water-based paint, dye, and shellac to simulate a wood-grain finish.
Film style A reference to the single-camera shooting method, typically identified with film practice, but also frequently employed in video production.
Fitting The process of adjusting a costume to fit the actor.
Flashback A cinematic technique whereby a reversal of the passage of time is indicated by some visual device, such as a whip pan or a rippling image effect.
Flat A framed, lightweight set unit that usually represents a portion of a wall.
Floor plan A scale drawing showing a studio or location from a bird's eye view.
Focus Either the optical property of attaining a sharp image in the lens or the psychological phenomenon of directing the audience's attention.
Fresnel A soft spotlight employing a Fresnel-type step-lens and fitted with barn doors to shape the beam of light.

Gaffer The chief electrician on a film crew.
Gaffer grip A special clamp that permits location lighting instruments to be attached to various architectural surfaces (such as window frames or picture moldings).
Gaffer tape A special cloth tape with a low-tack adhesive that permits things (such as cables) to be attached to floors or walls without damaging the surface.
Gate In lighting, this is the focal point inside the instrument (usually a Leko) where the shutters and gobos (cookies) are placed to give a sharp edge to the beam of light or pattern being projected.
Gobo (also cucaloris) A pattern projection device cut from a metal plate and inserted into a Leko.
Grab The process of freezing an image electronically.
Graphic Any static artwork used in shot; this could be an electronically produced illustration or a painted logo on a flat or a hard piece of artwork put in front of a camera.
Grid The metal structure in the ceiling of a studio from which lighting equipment and set pieces may be hung.

Hand-off (also toss) When one head in shot verbally passes the focus of attention to another head.
Hanger The hardware by which a lighting instrument is hung from the studio grid.
Hard graphic A piece of artwork executed in the old-fashioned way on a

piece of card stock and inserted into the video system by pointing a camera at it.

Hard-wired When a circuit is more or less permanently wired (through a metal conduit, for instance) and not intended to be easily disconnected or reconfigured, it is said to be hard-wired.

HDTV (high-definition (or high-density) television) One of a number of increased-resolution wide-screen video formats.

Head A term used in several contexts: The electronic core of a camera may be called the head, as may the camera mount on a tripod or pedestal. It also may refer to a person in shot, as in talking head.

High key A lighting term referring to exceptionally intense (brightly lit) scenes.

Independent scenery Any scenic unit that can stand alone—which is not structurally dependent on anything else.

Interactivity The concept of audience access to and input into the television system. True interactivity is antithetical to commercial broadcast TV as it exists today.

Jack A type of scenery brace, usually in the shape of a triangle, used to support a flat or a run of flats.

Key In lighting, an instrument that is the principal and most intense source in a given setup. In video imaging, the effect produced by superimposing one image on another.

Key and fill The lighting system that enhances the three-dimensionality of an in-shot head by means of unequal intensity.

Keyer The device that performs a video key effect.

Kicker An extra lighting source used to add sharpness to a head or object in shot.

Landing In scenic and architectural use, the platform either at the top of a flight of stairs or in the middle, but significantly larger than the stair treads.

Leko A lighting instrument using plano-convex lenses, that gives a very tightly controlled hard beam.

Luminance The property of brightness, as a function of the video imaging system.

Luminance key A key effect that favors the brighter image over the darker; a precursor to chroma-key.

Neutral density A lighting gel (filter) used to reduce the output of a lighting instrument without altering its color.

Newel post The large, often ornate, post at the bottom of a stairway or at the end of a balustrade.

Noise In video terminology, unwanted picture disturbances that usually occur in the dark areas of low-contrast images.

NTSC The system of video imaging established in the United States in the late 1940s.

Open-face spotlight A lighting instrument, typically used on location, which has no lens but gives a fairly hard light and is fitted with barn doors like a Fresnel.

PAL The German video imaging system, now adopted in most world markets.
Perspective The apparent convergence of parallel lines over distance—an optical illusion by which the brain judges the relative distance of things.
Pigment Chemical substances that give paint or dye its color properties.
Pilaster A false column, flattened out, which may be used as a wall decoration and is a fine device to cover joints in a run of flats.
Pixel One of the tiny dots on a TV screen that makes an image.
Primary colors The set of three colors which (theoretically) can be combined to make any other color. In pigment, these are red, yellow, and blue; in light, they are magenta, green, and cyan.
Profile piece A piece of timber (often plywood) cut in the silhouette of a desired shape and arranged in a row—like vertebrae—and covered with cardboard or canvas to make a tree or a large molding, such as a cornice.
Prosthetic A term referring to makeup that profoundly alters the shape of the face, such as a latex mask.

Rendering The process of putting one's design ideas into concrete form to communicate them to others. Also refers to the resultant artwork.
Riser The vertical face of a step in a stair unit.
Roller The (archaic) term for a scrolling credit list, derived from the fact that it used to be a hard graphic on a long strip of pliable card that actually rolled past the camera lens.
Rostrum camera A camera in fixed position over a light table, used to input hard graphics into a video graphic system.

Sash window A window in two sections, which slide vertically in tracks.
Scoop A lensless floodlight used in studio to cover large areas or where a very soft light is desired.
Seamless A term used in a number of contexts (most often in editing) that denotes a perfectly smooth transition.
Section A scaled drawing showing a unit from the side as though it had been cut through, in order to illustrate details of interior construction.
Shot plotter A drafting tool, something like a protractor, that permits the designer to plot shot angles in plan and section and thus construct an accurate sketch of the shot.
Site plan The location equivalent to a studio floor plan; the site plan may not be as accurately scaled.
Soft focus In lighting, the adjustment of the beam of light so as to remove the hard edge. In camera usage, the application of a special filter to give a dreamlike effect.
Specular reflection The sharp reflection of light such as is given by chrome-plated or other very shiny surfaces.
Stage Brace The telescoping brace that attaches to the back of a flat and to the floor to hold the flat up.
Storyboard A series of drawings to illustrate a sequence of shots.

Strike   In theatre, to disassemble the set and lights and load them out of the theatre or studio.

Stringer   In a stair unit, the sawtoothed piece of timber that holds the whole unit together.

Strip light   A lighting unit that consists of a series of boxes, each with a lamp and a color frame. Typically used to light large expanses of scenery such as backcloths.

Talent   The term used to refer to performers, news readers, or anyone paid to appear on camera.

Thickness   (also reveal) A false depth effect built onto scenic flats to simulate the thickness of a real wall.

Tracking shot   A shot in which the camera moves along with the action.

Tread   The horizontal part of a step on which people tread as they walk up or down (hence, the name).

Trompe L'Oeil   Literally, "deceptive appearance:" a hyper-realistic style of scene painting.

Truck   A camera movement term referring to any lateral motion across the shot (left-to-right, for example).

Truss   A metal structure of great strength, which is commonly used to support lighting equipment on location (typically at rock concerts) suspended from a building structure or towers constructed for the purpose. Large scenic units may also be supported from the floor on truss structures.

Vectorscope   A device in the video system that gives a graphic reading of chrominance.

Vehicle   The part of paint or dye that carries the pigment to the surface to be painted. For scenic paint, the vehicle is usually water.

Waveform monitor   A device that gives a graphic reading of luminance in the video system.

Whip-pan   A sudden and very rapid lateral movement of the camera.

Yoke   The part of a lighting instrument that is connected to the hanger or clamp; typically a metal strap in the shape of an upside-down U.

Zoom   The optical effect that makes the subject of a shot appear to move toward the camera or away from it and is accomplished by movement of elements within the lens.

Recommended Reading

Books

Design Resources
Praz, Mario. *An Illustrated History of Furnishing.* New York: Braziller, 1964.
Speltz, Alexander. *The Styles of Ornament.* New York: Dover.

Scenery
Bowman, Ned. *Handbook of Technical Practice for the Performing Arts.* Norwalk, CT: Scenographic Media, 1975.
Burris-Meyer, Harold, and Edward C. Cole. *Scenery for the Theatre.* Boston: Little Brown, 1972.
Gillette, A.S. *Stage Scenery: Its Construction and Rigging,* 3rd ed. New York: Harper Collins, 1981.
Parker, W. Oren, Harvey K. Smith, and Craig Wolf. *Scene Design and Stage Lighting,* 6th ed. New York: Holt, Rinehart, Winston, 1990.
Pecktal, Lynn. *Designing and Painting for the Theatre.* New York: Holt, Rinehart, 1975.

Lighting
LeTourneau, Tom. *Lighting Techniques for Video Production: The Art of Casting Shadows.* White Plains, New York: Knowledge Industry Publications, 1987.
Mathias, Harry and R. Patterson. *Electronic Cinematography.* Belmont, CA: Wadsworth, 1985.
McCandless, Stanley, *A Method of Lighting the Stage.* New York: Theatre Arts Books, 1958.
McGrath, Ian. *A Process for Lighting the Stage.* Needham Heights, MA: Allyn and Bacon, 1990.
Millerson, Gerald. *Technique of Lighting for Television and Motion Pictures.* Stoneham, MA: Focal Press, 1982.
Pilbrow, Richard. *Stage Lighting.* New York: Drama Book Publishers, 1991.
Viera, Dave. *Lighting for Film and Electronic Cinematography.* Belmont, CA: Wadsworth, 1993.

Properties

Bryson, Nicholas L. *Thermoplastic Scenery for the Theatre.* New York: Drama Book Specialists, 1972.

Kenton, Warren. *Stage Properties and How to Make Them.* New York: Drama Book Specialists, 1978.

Costumes

Anderson, Barbara B. and Cletus. *Costume Design.* New York: Holt, Rinehart, Winston, 1984.

Barton, Lucy. *Historic Costume for the Stage.* Baker's Plays, 1961. (Out of print).

Waugh, Norah. *The Cut of Men's Clothes: Sixteen Hundred to Nineteen Hundred.* New York: Theatre Arts Books, 1964.

___. *The Cut of Women's Clothes: Sixteen Hundred to Nineteen Thirty.* New York: Theatre Arts Books, 1968.

Makeup

Corson, Richard. *Stage Makeup,* 8th ed. New York: Prentice Hall, 1986.

Kehoe, Vincent. *Technique of the Professional Makeup Artist.* Stoneham, MA: Focal Press, 1985.

Smith, Dick. *Dick Smith's Monster Makeup Book.* Imagine Press, 1985.

Video Practice

Armer, Alan A. *Directing Television and Film.* Belmont, CA: Wadsworth Publishing Co., 1988.

Zettl, Herbert. Sight–Sound–Motion: Applied Media Aesthetics. Belmont, CA: Wadsworth Publishing Co., 1990.

——. *Television Production Handbook,* 5th ed. Belmont, CA: Wadsworth Publishing Co., 1992.

Addendum

The preceding list offers a good starting point for a reference library. As you acquire projects, you will also acquire resource materials specific to them. Large format art books—sometimes referred to as table-top or coffee table books— are often invaluable resources on style, color, and details of props and costumes; it will serve you well to browse among the sale bins in book stores as you begin to build your reference collection.

Periodicals

Creative Review

A good monthly magazine, focusing on commercial design, graphics, film, and video. Published in the United Kingdom, but international in scope.

Lighting Dimensions

A trade magazine for all aspects of the lighting industry: stage, film, video, architectural.

Theatre Crafts
A trade magazine for theatre technicians and designers. Especially good for novices and students.

Videography
Another trade magazine, focusing on the video industry, especially non-broadcast, and very good for technical information.

Index

acrylic, 139
    paint, 142
actors, 6–7
Advanced Television (ATV), 49
advertising, 10, 24
    and design, 25–27
    and lighting, 26
"Alf," 74–75
"Alien Nation," 74–75
Altman, Robert, 19
analine dye, 143
angle of acceptance, 113
animation
    and computer, 60, 82–84, 174
    and fantasy, 77, 81–84
architecture, 13–14
art director, 17
artifact, 181
aspect ratio, 31–32, 181

backflap hinge, 181
backlighting, 158–159, 169, 181
Ball, Lucille, 9
balusters, 136
balustrade, 136, 181
barn door, 163–164, 181
baseboard, 132, 181
BBC, 45, 84, 141, 179
Bel Geddes, Norman, 13
Berle, Milton, 9
Bibiena family, 13
binder, 142, 181
Blade Runner, 74
blocking, 44–45
"Booker," 25
Boot, Das, 47
Bosch DaVinci, 79

Bowman, Ned, 144
brace, 181

cable television, 176–177
CAD, 116–117
camera
    and color, 12–13, 48–51
    film, 57
    and focus, 43
    lens, 43, 57, 77–78, 93
    moving, 45–46
    and news program, 93–95
    plotting, 92–93
    and reality, distortion of, 68
Candide, 7
cap gen, 171–172, 181
Capra, Frank, 77
captioning, 171–172
car commercials, 26–27
Cartwright, Jim, 45
casement, 131, 181
casing, 132, 181
cel, 82, 181
centerline, 120, 181
chair rail, 132, 181
character generators, 59
"Cheers," 41–43
"Cheyenne," 10
"China Beach," 10
chip, 57, 181
chroma-key unit, 59, 79–81, 182
    and weather, 59, 66, 81
chrominance, 181
cinema. *See* film
Cinerama, 32
Colgate-Palmolive, 9
color, 48–55

and camera, 12–13, 48–51
complementary, 160, 182
designing in, 49–51
hue, 50
monitoring, 12–13
and news program, 52–53, 67–68
primary, 185
and production designer, 49–51
psychology of, 52–54
saturation, 50, 53
and talk show, 54
color bars, 49, 154, 182
color-separation overlay (CSO), 59
color wheel, 182
Come Back to the Five and Dime, Jimmy Dean, 19
comet tail, 50
Commedia dell'Arte, 3–4
commercial role of design, 24–25
commercials, 12, 25–27, 29
and focus, 43
and storyboards, 91–92
composition, 31–36
film, 6–7
formulas, 6
and multiple-head shots, 34–36
and news program, 32–36
and talk show, 34–36
computer, 59–61
and animation, 60, 82–84, 174
and design, 116–117
and editing, 60
and floor plan, 117
and graphics, 82–84, 171, 173–174
network, 176–177
Connery, Sean, 31–32
continuity, 19–20
and production designer, 63
and set dressing, 145
contrast ratio, 49, 52, 182
control monitor, 12–13
cornice, 132, 182
corporate teleconference, 178
costume designer, 20–21, 147
and director, 21
and hair, 20
and makeup, 20, 148, 152
and news program, 20–21
and production designer, 18, 148–150, 152–153
costume plot, 20, 149–150
CSO, 59, 182
cube wipe, 59

cucaloris, 164, 182
Curtin, Jane, 33
cut, 46
cyc light, 182
cyclorama, 103, 163, 182

"Dallas," 76
Death of a Salesman, 89
degradation of image, 82, 182
dentil, 132
dependent scenery, 182
detail, 13, 108–109, 182
diazo, 115
diffusion material, 165, 182
digital technology, 61, 81
and animation, 84
dimensioned drawing, 182
dimension line, 121–123
dimmers, 162–163, 182
dip to black, 75–77, 182
director, 28, 62
and costume designer, 21
and focus, 41
and news program, 66–67
and production designer, 36, 41, 71–72
and script, 89
Disney, Walt, 48
"Doctor Who," 77
doors, 106–107, 128–130
downstream, 78, 182
drafting technique, 115–124
language, 118–123
media, 115–117
style, 123–124
drawings, 14–15, 98–99, 108–11
ink, 115–116
pencil, 115–116
dressing light, 35, 159–161
dynamics, 44–47

Eastenders, 137
editing, 11, 56
and computer, 60
and digital technology, 61
and production designer, 63
electronic imaging, advantages, 58–59
elevation, 108–109, 123–124, 183
European television, 11, 17
eyeline, 32–33, 183

fantasy, 73–84
and animation, 77, 81–84

mechanics of, 74–81
Farquharson, Alan, 71
fiberglass, 137
film
    actors, 6–7
    camera, 57
    formats, 32
    lighting, 110–111
    mechanics, 56
    and scenery, 5–6
    and television design, 8–11
    and theatre design, 6–7
film style, 41, 183
filters, 77–78, 163
Fisher, Jules, 48
fitting, 183
"Flash, The," 81
flashback, 76–77, 183
Flatliners, 76
flats, 18, 127–133, 183
floodlights, 157
floor director, 17
floor plan, 92–93, 106–110, 183
    and computer, 117
    and rendering, 100
focal plane, 57
focus, 37–43, 183
    and camera, 43
    directing, 37–41
    and news program, 38–41, 43
footcandle, 157
Ford, Harrison, 31–32
Fox, Michael J., 10
Fox Network, 25
French brace, 128
French enamel varnish (FEV), 143, 183
Fresnel light, 159, 163–164, 183
Friendly, Fred, 17
furniture, 139–140
    and model, 105

gaffer, 183
gaffer grip, 165, 183
gaffer tape, 183
game shows, 46
    lighting, 51
gate, 164, 183
gels, 163
glare, 146
Gleason, Jackie, 17
gobo, 164, 183
gouache painting, 55, 103–104
grabbing images, 84, 183

graphics, 171–175, 183
    aesthetics, 174–175
    artwork, 172–174
    and camera shots, 93–95
    and captioning, 171–172
    and computer, 82–84, 171, 173–174
    and fantasy, 75–76
    hard, 183
    and news program, 33–34, 93–95
    personnel, 172
    and production designer, 18
    and special effects, 59
    and weather, 59, 174
Greek theatre, 3–4
grid, 111–112, 183
Grogan, Mick, 140
"Gunsmoke," 11

hair and costume designer, 20
hand-off, 40–41, 183
hanger, 163, 183
hard flat, 127–128
hard-wired circuit, 184
head, 57
high-definition television (HDTV), 11, 15–16, 49, 177
    aspect ratio, 32
    and color, 51
"Hill Street Blues," 10–11, 25, 27
Honda commercials, 27
"Honeymooners, The," 9, 180

Ianiro lights, 164–165
identification (ID) block, 123
image of station, 24–25
incognito shot, 43
"Incredible Hulk, The," 76
independent door, 129
independent scenery, 184
Indiana Jones and the Last Crusade, 31–32
industrial shows, 12
interactive television, 176–177, 184
Intergalactic Serial Shop Cookbook, The (Bowman), 144
international television format, 10–11
interpretation, 64–72
    degrees of, 73–74
It's A Wonderful Life, 77

jacks, 128, 184
"Jeopardy," 51

jogs, 133
Jones, Inigo, 13

key and fill lighting, 157–159, 169, 184
keyer, 184
kicker, 157, 172, 184

landing, 184
Lathe of Heaven, The, 74–75
Lee, Eugene, 7, 30
LeGuin, Ursula, 74
Leko, 164, 184
letterboxing, 32
lighting, 13, 154–170
    and advertising, 26
    aesthetics, 166–170
    back, 158–159, 169, 181
    and color, 51
    dimmers, 162–163, 182
    dressing, 35, 159–161
    and film, 110–111
    focusing, 22
    and game shows, 51
    key and fill, 157–159, 169, 184
    levels, 12
    and locations shoots, 22
    mechanics, 161–166
    and news program, 35, 52
    origins of, 3–4
    planning, 22
    plot, 22, 110–114
    principles of, 156–157
    and production designer, 18, 51–52, 70–71
    rigging, 22–23
    and shot plot, 113–114
    and soaps, 43, 51
    and talking heads, 43
    theatre, 3–4
    training, 11–12
    and walls, 169
lighting designer, 17, 22–23, 154–156
    and production designer, 156, 169–170
light meter, 157
light plot, 167–169
    drafting, 118–119
line weight, 115, 118–119, 123
Little Big Man, 152
location shoots
    and lighting, 22
    and models, 102
    and rendering, 101–102
    and set dressing, 145–146
    site plan, 107, 118
logos, 32–34, 94
Lucas, George, 7, 74
lumens, 157
luminance key, 81, 171, 184
Lynch, David, 76

McLane, Derek, 70–71
makeup, 147–148, 150–152
    and costume designer, 20, 148, 152
    and news program, 151
"Married With Children," 9, 25
"M*A*S*H," 11
matte, traveling, 7
"Maverick," 11
Mercedes-Benz, 26
metal and set construction, 137
"Miami Vice," 25
Mielziner, Jo, 88–89
models, 67, 100–101, 104–105
    and location shoot, 102
MTV, 78
multiple cameras, 9, 41–43, 111
multiple-head shots, 34–36
"My Favorite Martian," 74–75

network
    computer, 176–177
    and image, 25
neutral density filters, 163, 184
newel post, 137, 184
news program, 8, 29
    camera shots, 93–95
    and color, 52–53, 67–68
    composition, 32–36
    and computer animation, 84
    and costume designer, 20–21
    design solution, 65–68
    and director, 66–67
    and field footage, 60
    and focus, 38–41, 43
    format, 64–65
    and graphics, 33–34, 93–95
    and lighting, 35, 52
    and makeup, 151
    and revenues, 24
noise, 184
Nolan Brothers' Studios, 141
nonbroadcast television, 178
nosing, 136
NTSC television format, 10–11, 48, 184

aspect ratio, 31
scan, 571

O and O, 24–25
objectives, 24–30
    clarity in, 27–30
oblique projection, 123–124
open-face spotlight, 185
opera, 9

Paintbox, 59–60, 173–174
painting, 50–51, 54–55
    and scenery construction, 141–144
PAL television format, 11, 41, 48, 185
    aspect ratio, 31
    scans, 57
panel moldings, 133
Pauley, Jane, 71
perspective, 38, 185
perspex, 131, 139
Petersen, Wolfgang, 47
picture rails, 132
pigment, 54–55, 141–142, 185
pilaster, 133, 185
pixelization, 78–79
    wipes, 58–59
plastics, 137–138, 140–141
    painting, 143–144
platforms, 106, 133–135
polyvinyl chloride (PVC), 139
Potter, Dennis, 75
preproduction, 28–29, 62, 72
presentation studio, 65
production designer, 62
    and blocking, 44–45
    and color, 49–51
    and continuity, 63
    and costume designer, 18, 148–150, 152–153
    defined, 1
    and director, 36, 41, 71–72
    and editing, 63
    and future, 15–16
    and graphics, 18
    and lighting, 18, 51–52, 70–71
    and lighting director, 156, 169–170
    origins of, 3–16
    and props, 18
    responsibilities, 18–20
    roles, 1–2
    staff, 17–18
    training, 15
profile piece, 185

prop master, 19
props, 19
    construction, 19
    and location shoots, 145–146
    and production designer, 18
    rental, 19
prosthetics, 151–152, 185

Quantel, 59–60

ratings, 24
"Real Life," 70–71
Renaissance theatre, 3–4
rendering, 14–15, 97–105, 185
    final, 102–105
    and models, 100–101, 104–105
    preliminary, 97–99
reveal, 129
riser, 135–136, 185
"Road," 45
Rojo, Jerry, 7
rostrum, 133
rostrum camera, 84, 185

Sajak, Pat, 29
sash window, 185
"Saturday Night Live," 29–30, 33
scaffolding, 137
scale, 121
scenery construction, 18–19, 125–139
    painting, 141–144
    special, 137–139
    3-D treatments, 140–141
Schneider, Frank, 1, 67–68, 71, 126
science fiction, 7, 18, 77
scoop, 165, 185
Scott, Ridley, 74
scout, 101
script, 85–89
    analysis, 89–91
SECAM, 31
section, 109, 185
section line, 120–121
Sennett, Mack, 3
"Sesame Street," 30
set dressing, 19, 144–146, 182
    and location shoot, 145–146
Shandling, Garry, 25
shot plot, 92–93
    and lighting, 113–114
    and location, 101–102
shot plotter, 93, 114, 185
sideband, 58

sight lines, 6, 12
"Simpsons, The," 82
"Singing Detective, The," 75
sitcoms, 9–10, 36
site plan, 107, 118, 185
Skelton, Red, 9
soaps, 9–10, 71
    lighting, 43, 51
soft focus, 76–77, 185
Sony Betacam, 60
Sony 1125 HDTV, 11, 16, 140, 176
special effects, 16, 59
specular reflection, 185
Spielberg, Steven, 31–32
spotlight, 157
    open-face, 185
stage brace, 128, 185
stairs, 135–137
star filter, 78
"Star Trek," 77
Star Wars, 7, 74
Stone, Jon, 30
storyboard, 62, 91–96, 99–100, 114, 185
    and focus, 41
    and location shoot, 101–102
strike, 133, 147, 186
stringer, 136, 186
strip light, 165–166, 186
Sugar, 143
Svoboda, Josef, 13–14, 175
switcher, 58

talent, 6, 186
talking heads, 32–33
    and aspect ratio, 31
    and camera shots, 94
    lighting, 43
talk show, 29, 69
    and color, 54
    and composition, 34–36
technical director, 17
telephone and television, 179
theatre
    actors, 6
    and film design, 6–7
    lighting, 3–4
    and production design, 11–13

    and television, 15
thickness, 129, 186
3-D imaging, 60–61, 83
time-base corrector, 78
"Tonight Show," 29
tracking shots, 45–46, 186
tread, 135–136, 186
Tremors, 38
trompe l'oeil, 126–127, 186
truck, 46, 186
truss, 186

unions, 18, 23

variety shows, 29
vectorscope, 49, 186
vehicle, 141–142, 186
videotape recording, 56–58, 82
vision mixer, 58
Volvo, 26

walls
    and lighting, 169
    textured, 140–141
    wild, 36, 128
wardrobe, 21, 147–149, 250
Warner Brothers, 8, 10
waveform monitor, 154, 186
weather
    and chroma-key unit, 66, 81
    and graphics, 59, 174
"Wheel of Fortune," 46, 53
whip-pan, 76, 186
white and camera, 50
wide shot, 95
wigs, 152
Wild At Heart, 76
wild wall, 36, 128
windows, 131–132
wipes, 58–59
Wizard of Oz, 48, 76

yoke, 163, 186

zero date, 123
zoom, 46, 93, 186